Autonomy and Independence

Aging in an Era of Technology

Synthesis Lectures on Technology and Health

Editors

Ronald M. Baecker, *University of Toronto*
Andrew Sixsmith, *Simon Fraser University and AGE-WELL NCE*
Sumi Helal, *University of Florida*
Gillian R. Hayes, *University of California, Irvine*

The series publishes state-of-the-art short books on transformative technologies for health, wellness, and independent living. Our scope of publishing in the expanding health tech field includes:

- Technology in support of active and healthy living and aging

- Digital technologies for health- and social-care improvement

- Diagnostic, screening, and tracking tools

- Assistive and rehabilitative technologies

The series includes a subseries of books published in partnership with Canada's AGE-WELL NCE that specifically addresses their 8 AgeTech Challenge Areas.

Each lecture introduces the context in which the technology is used—wellness, health, medicine, special needs, or other contexts. Authors present and explain the technology and review promising applications and opportunities as well as limitations and challenges. They include material on their own work while surveying the broader landscape of related research, development, and impact.

Autonomy and Independence: Aging in an Era of Technology
Lili Liu, Christine Daum, Antonio Miguel Cruz, Noelannah Neubauer, and Adriana Ríos Rincón

Brain–Computer Interfaces: Neurorehabilitation of Voluntary Movement after Stroke and Spinal Cord Injury
Cesar Marquez-Chin, Naaz Kapadia-Desai, and Sukhvinder Kalsi-Ryan

Ending Medicine's Chronic Dysfunction: Tools and Standards for Medical Decision Making
Lawrence L. Weed (1923–2017) and Lincoln Weed

Research Advances in ADHD and Technology
Franceli L. Cibrian, Gillian R. Hayes, and Kimberley D. Lakes

Design and the Digital Divide: Insights from 40 Years in Computer Support for Older and Disabled People
Alan F. Newell

AGE-WELL NCE Inc. (www.agewell-nce.ca) is Canada's Technology and Aging Network. The pan-Canadian network brings together researchers, older adults, caregivers, partner organizations, and future leaders to accelerate the delivery of technology-based solutions that make a meaningful difference in the lives of Canadians. AGE-WELL researchers are producing technologies, services, policies, and practices that improve quality of life for older adults and caregivers and generate social and economic benefits for Canada. AGE-WELL is funded through the Government of Canada's Networks of Centres of Excellence (NCE) program.

The STAR (Science and Technology for Aging Research) Institute (www.sfu.ca/starinstitute) at Simon Fraser University (SFU) is committed to supporting community-engaged research in the rapidly growing area of technology and aging. The Institute supports the development and implementation of technologies to address many of the health challenges encountered in old age, as well as addresses the social, commercial, and policy aspects of using and accessing technologies. STAR also supports the AGE-WELL network.

Autonomy and Independence: Aging in an Era of Technology
Lili Liu, Christine Daum, Antonio Miguel Cruz, Noelannah Neubauer, and Adriana Ríos Rincón

ISBN: 978-3-031-03754-2 Paperback
ISBN: 978-3-031-03764-1 PDF
ISBN: 978-3-031-03774-0 Hardcover

DOI: 10.1007/978-3-031-03764-1

A Publication in the Springer series
SYNTHESIS LECTURES ON TECHNOLOGY AND HEALTH
Lecture #19
Series Editors: Ron Baecker, University of Toronto, Andrew Sixsmith, Simon Fraser University and AGE-WELL NCE, Sumi Helal, University of Florida, and Gillian R. Hayes, University of California, Irvine

Series ISSN PENDING Print PENDING Electronic

Autonomy and Independence

Aging in an Era of Technology

Lili Liu
University of Waterloo, Canada

Christine Daum
University of Waterloo, Canada

Antonio Miguel Cruz
University of Alberta, and Glenrose Rehabilitation Hospital, Canada

Noelannah Neubauer
University of Waterloo, Canada

Adriana Ríos Rincón
University of Alberta, Canada

SYNTHESIS LECTURES ON TECHNOLOGY AND HEALTH #19

ABSTRACT

This book looks at how AgeTech can support the autonomy and independence of people as they grow older. The authors challenge readers to reflect on the concepts of autonomy and independence not as absolutes but as experiences situated within older adults' social connections and environments. Eleven personas of people around the world provide the context for readers to consider the influence of culture and values on how we understand autonomy and independence and the potential role of technology-based supports.

The global pandemic provides a backdrop for the unprecedentedly rapid adoption of AgeTech, such as information and communication technologies or mobile applications that benefit older adults. Each persona in the book demonstrates the opportunity for AgeTech to facilitate autonomy and independence in supporting one's identity, decision making, advance care planning, self care, health management, economic and social participation, enjoyment and self fulfillment and mobility in the community. The book features AgeTech from around the world to provide examples of commercially available products as well as research and development within the field. Despite the promise of AgeTech, the book highlights the "digital divide," where some older people experience inadequate access to technology due to their geographic location, socio-economic status, and age.

This book is accessible and relevant to everyday readers. Older adults will recognize themselves or peers in the personas and may glean insight from the solutions. Care partners and service providers will identify with the challenges of the personas. AgeTech entrepreneurs, especially "seniorpreneurs," will appreciate that their endeavours represent a growing trend. Researchers will be reminded that the most important research questions are those that will enhance the quality of life of older adults and their sense of autonomy and independence, or relational autonomy and interdependence.

KEYWORDS

Autonomy, Independence, Aging, Technology, AgeTech

Contents

Acknowledgments

The authors thank the individuals featured as real-life personas in this book: Marni Panas, Roger Marple, Gladys Rincón Neira, Edmund Bauer, Jamie Stirling, Enma Cruz Delgado, and Elder Leslie (Les) Nelson. We are grateful for their willingness to share their experiences, stories, and time with us. We also thank Nicole Bird and Susie O'Shea for preparing the persona for Elder Les; Adebusola Adekoya for authoring the persona for Titi Torike; and Drs. Masako Miyazaki and Toshio Ohyanagi for contributing the persona of Kazuko Hasegawa. We are privileged to learn from them.

We thank Hector Perez (Postdoctoral fellow, University of Waterloo), Adebusola Adekoya (Ph.D. student, University of Waterloo), Lauren McLennan (Undergraduate student, University of Waterloo), and Melika Torabgar (Master's student, University of Alberta) for their contributions to text and figures. Hector Perez's design skills and creativity were integral in the development of many of the images contained in this book. Esha Kulkarni (Undergraduate student, University of Waterloo) is the talented illustrator and designer who created the drawings and snapshots of the personas that appear in each part.

The authors acknowledge the exceptional team of research assistants for their unwavering support and commitment to working on this book including Serrina Philip, Elyse Letts, Emily Rutledge, Sharanya Thamilvaanan, and William Tang. We thank Christine Thelker and Burn Evans for sharing their experiential knowledge and for allowing us to use their pictures in this book.

We appreciate Andrew Sixsmith for his mentorship and critique of the book throughout its evolution as well as Juliet Neun-Hornick for her organizational skills. We would like to thank Christine Kiilerich and Diane D. Cerra of Morgan & Claypool Publishers for their essential role in bringing this book to fruition. We acknowledge AGE-WELL for the financial support. The authors would like to thank everyone in the AGE-WELL community—Network Management Office, researchers, older adults and caregivers, and partners. Their passion for innovation was essential in developing and sustaining this book's creation. Thank you for your collaborative spirit.

Abbreviations

ADL	Activity of Daily Living
AGM	Annual General Meeting
ARIA	Alberta Rating Index for Apps
CMOP-E	Canadian Model of Occupational Performance and Engagement
COPD	Chronic Obstructive Pulmonary Disease
CRPD	Convention on the Rights of Persons with Disabilities
GATE	Global Cooperation on Assistive Technology
GPS	Global Positioning System
HoT	Home of Things
IADL	Instrumental Activity of Daily Living
ICF	The International Classification of Functioning, Disability, and Health
ICMOP-E	Integrated Canadian Model of Occupational Performance and Engagement
ICT	Information and Communication Technology
IoT	Internet of Things
LGBTQIA2	Lesbian, Gay, Bisexual, Trans, Queer, Intersex, Asexual, and Two-Spirit
mHealth	Mobile Health
MoCA	Montreal Cognitive Assessment
MOOC	Massive Open Online Course
MMSE	Mini-Mental State Examination
RFID	Radio Frequency Identification
SSI	Self-Sovereign Identity
TAGlab	Technologies for Aging Gracefully Lab
U3A	University of the Third Age

Introduction

It is projected that between 2015 and 2050, the number of people aged 60 years or older will double, reaching 2.1 billion or 22% (United Nations, 2021). According to the World Health Organization (2020), over one billion people live with a disability. Disabilities are higher among older adults, and about 46% of older adults live with conditions that prevent activity performance (United Nations, 2021). In 2017, the prevalence of disabilities among older Canadians was slightly higher in women then in men in the 65–74 age group (33.3% vs31.5%) and in the 75 years and older group (49.2% vs. 45%) (Statistics Canada, 2018a). The most frequent types of disabilities among older Canadians are pain-related, mobility, flexibility, hearing, dexterity, and sight (Statistics Canada, 2018c), many of which are related to chronic conditions associated with aging. Among Canadians aged 60 years or older, 15% experience mental disorders, the most common of which is dementia and depression which affect 5% and 7% of the world's older adults (World Health Organization, 2017). Disabilities can diminish individuals' autonomy and independence (DiGennaro Reed et al., 2014).

The nexus of global aging, a worldwide pandemic, the associated exponential growth in technology adoption, and our heightened awareness of inequities situates the context for this book. This book is not a product repository. Instead, it highlights some key issues and showcases some technology examples pertaining to autonomy and independence. We present 11 personas, varying in amount of detail, to establish the context of autonomy and independence and the roles of technology. Most personas are older adults, while two are aging and preparing for older adulthood. Some personas are derived from interviews, some are based on actual people, and others are based on a combination of people.

While the book includes personas from around the world, the book is written from a Canadian perspective, and reflects issues currently relevant to the Canadian context, such as equity, diversity, and inclusion. We challenge readers to consider these issues in their own contexts. The authors' disciplinary orientation is that of health. Although the personas present with a range of health conditions, the book tends to refer to dementia or age-related cognitive decline which is a focus of the authors' research programs.

Technologies can support older adults' autonomy and independence (Sixsmith et al., 2020). As in the first book of this series, *AgeTech, Cognitive Health, and Dementia* (Sixsmith et al., 2020), this book uses the term "**AgeTech**" to refer to the use of technologies, such as information and communication technologies (ICTs), that benefit older adults (Pruchno, 2019). AGE-WELL prefers this terminology over "*gerontechnology*" which is not as widely used beyond Europe and academia.

This book is organized into four parts: (I) Technology for Autonomy and Independence: An Overview, (II) How can Technology Support One's Autonomy?, (III) How can Technology Support One's Independence?, and (IV) Challenges and Future Directions.

In Part I, we examine the concepts of autonomy and independence (Chapter 1). In the context of this book, autonomy refers to an ability to make decisions for oneself, whereas independence is the degree to which a person can perform a task or activity. However, we challenge readers to consider the definitions of autonomy and independence, not as absolutes and from individual perspectives, but as evolving concepts that consider older adults' social connections and environments. We provide examples that persuade us to reconceptualize these two terms more as "relational autonomy" and "interdependence" to reduce the stigma associated with aging and disability. We review international frameworks on health and technologies (Chapter 2) and how technologies help older adults enjoy basic human rights. The 2006 United Nations Convention on Rights of Persons with Disabilities advocated that access to technologies, including the Internet, and nations are obligated to develop strategies to promote the development and commercialization of affordable technologies so that they are available for individuals who need them. In this Part, we use a persona to illustrate that some members of under-represented LGBTQIA2 communities feel a need to plan for their aging earlier than others. A second persona is a person living with dementia and uses a GPS device to maintain her independence in walking while giving her husband peace of mind. The third persona is a once-independent and socially active woman living in South America, who became restricted in her independence and activity engagement during the COVID-19 pandemic because she does not have access to the Internet.

Part II consists of five chapters (3–7) and covers topics related to one's sense of autonomy: sense of self, capacity, advance care planning, risk, and privacy. This part begins with a persona of a man living with dementia, by himself, who embraced technology earlier in his dementia journey and continues to use AgeTech in his everyday activities as he plans for his future and weighs the risks. In each chapter, we describe the underlying topic, its role on autonomy of older adults, and ethical tensions. Chapter 3 examines sense of self or one's identity and how digital storytelling, information communication technologies, and assistive technologies can help persons with declining cognitive function to preserve their sense of identity. In Chapter 4, we present capacity in the context of decision making. When traditional assessments of capacity are not feasible or practical due to diminished attention, innovations such as serious games may be used as proxies for cognitive assessment. Advance care planning (Chapter 5) examines the importance of planning ahead for when one is not able to make decisions, including whether one wishes to be monitored with technologies. We foresee the application of Self-Sovereign Identity with older adults who own their personal data and plan ahead, and under what circumstances, this data could be used to ensure they can live in their communities safely. Chapter 6 discusses the concept of risk, what is perceived risk, and the balance between acceptable risk and autonomy. Finally, in Chapter 7, we present privacy as one's freedom to decide whether to disclose one's personal data, as an expression of one's autonomy. We also emphasize the importance of representing older adults in research on privacy concerns.

In Part III (Chapters 8–13), we discuss the role of AgeTech to facilitate independence in: Self Care, Health Management, Economic and Social Participation, Enjoyment and Self-Fulfillment, Mobility in the Community, and Usability of Technologies. In Chapter 8, we demonstrate how performance in daily activities such as feeding, bathing, cooking, and medication management can be facilitated by a variety of AgeTech including the Internet of Things (IoT) and Smart Homes. Health management (Chapter 9) is facilitated by everyday digital technologies particularly during the global COVID-19 pandemic when physiological and activity monitoring has been used as an alternative to in-person visits with health professionals. In Chapter 10, we explore economic and social participation and how AgeTech can facilitate older adults' contribution to their communities and society. These activities include paid work, volunteer services, caregiving, and civic participation, all of which can be enhanced by technologies. Chapter 11 examines enjoyment and self-fulfillment and ways that AgeTech allow or enhance engagement in spirituality, fun, pleasure, relaxation, learning, and growth, including third-age learning. In Chapter 12, mobility in one's community is discussed along with the role of tracking devices, community alert systems, and autonomous vehicles. The overall concept of usability is described in Chapter 13 with a real-life persona of an older adult who immigrated to the United States from Cuba. Despite being exposed to technologies for the first time when she was 68 years old, she displays characteristics of an "early adopter" and enjoys usable devices that are well-designed.

Part IV completes this book with Chapter 14 in which we present four challenges: (1) the meaning of autonomy and independence; (2) society inequities and the digital divide; (3) evolution of technologies; and (4) the importance of policies to make changes happen and to sustain them. We use three personas to bring these challenges to life: an Indigenous Elder who conducts ceremonies online during the pandemic, an African older woman who lacks access to the Internet, and a frail, Japanese elderly woman who relies on a robotic plush toy, enabled by artificial intelligence, to connect with her home care service providers. We conclude by recommending future directions to address these four challenges.

This book focuses on one of AGE-WELL's eight Challenge Areas, which are: (1) Supportive Homes & Communities, (2) Health Care & Health Service Delivery, (3) Autonomy & Independence, (4) Cognitive Health & Dementia, (5) Mobility & Transportation, (6) Healthy Lifestyles & Wellness, (7) Staying Connected, and (8) Financial Wellness & Employment. The concept of autonomy and independence intersects with the first book titled, *AgeTech, Cognitive Health, and Dementia*, and touches on the other challenge areas. Future books are being planned in this series that will examine other challenge areas in greater detail.

PART I

Technology for Autonomy and Independence: An Overview

"My lack of independence shouldn't impact my sense of autonomy"

Marni Panas

PI.1 THE CHALLENGE

Throughout this book, the idea of a "challenge" is a key component. Each challenge tries to be more than a description of a specific problem. This book tries to encompass opportunities, complexities, and innovation that apply to multiple problems. This first persona challenges us to examine the concepts of autonomy and independence from an aging process. Readers are challenged to consider that some segments of our populations must plan for their older adulthood earlier than others because of their gender identities.

Autonomy and independence are sometimes used interchangeably. The degree to which a person is "independent" is commonly used in Western cultures by health professionals in their assessments and intervention goals. Does the use of the term "independence" bias service providers and set up unrealistic expectations for service recipients? Can a person be dependent for certain tasks, but still be autonomous or in control of their own decisions? Can two persons with the same health care needs differ in their sense of autonomy and levels of independence? How does interdependence relate to independence? How does one's sense of identity relate to autonomy and independence?

This Part addresses these questions by examining the concepts of autonomy and independence in the context of aging. We relate these concepts to one's sense of identity. We also consider autonomy and independence in the context of aging and disability, gender, and culture in an era of technology.

PI.2 WHAT IS IN THIS PART?

In this part, we present two introductory chapters to discuss how technology supports the autonomy and independence of people as they age.

In Chapter 1, we introduce the concepts of autonomy and independence in the context of aging in an era of technology. In Chapter 2, we discuss relevant international frameworks on health and technology. We also discuss how access to some technologies is considered a human right. We introduce Marni Panas, who enriches the discussion about autonomy, independence, and technology from her lived experience in a world in which we need to keep advocating for inclusion and equity. While it may not seem readily apparent because Marni is relatively young, her persona and scenario remind the mainstream population that some under-represented segments of our society are compelled to plan for their aged years earlier to ensure they are included and treated equitably in their older adult years.

PERSONA, SCENARIO, AND SOLUTION

Marni Panas

- **Age:** 50
- **Gender:** Woman, pronouns are she/her
- **Lives in:** Single family home with partner, son, in Edmonton, Alberta
- **Social circle:** A few close friends, colleagues, broad online community

Signature interests

Parent, partner, runner, diversity and inclusion expert at a health service provider, Canadian Certified Inclusion Professional (CCIP), human rights advocate.

" My lack of independence shouldn't impact my sense of autonomy"

Health

Currently healthy but aware of need to prepare for her senior years.

Technology

Social media for information activism, safe online spaces for connecting within groups, access to services and information, databases that promote equitable representation of lesbian, gay, bisexual, trans, queer, intersex, asexual, and two-spirit (LGBTQIA2) communities.

Persona and scenario. Marni is a human rights activist and a transgender woman. She leads diversity and inclusion initiatives with a provincial health service. When she was planning for her advanced care directive, she was faced with the reality that society is not designed to make decisions that would respect her autonomy. Therefore, she asserted her autonomy by clearly documenting that if and when she is no longer able to make decisions about her own care, a specified agent would be contacted, not members of her birth family. She did not want to be in a position where others decide what to do for her based on the appearance of her body. Therefore, she chose to have gender affirming surgery—a decision based on her experiences and goals and would not be the same for any other person. This would mitigate decisions about her living environment and the type of care she receives as she ages. According to Marni, "If I couldn't live in peace, I want to be able to rest in peace." She emphasizes that each person's journey is unique and valid.

In her professional role as a manager of diversity and inclusion, Marni has advocated for older adults with an intersectional lens including race, gender identities, education, geography, and socio-economic factors. Like her, her clients use information and communication technology to connect with and share support among others with similar gender identities. This was particularly relevant during the SARS-CoV-2, hereafter referred to as the COVID-19, pandemic. The opportunity to work from home through technology platforms affords some lesbian, gay, bisexual, trans, queer, intersex, asexual, and two-spirit (LGBTQIA2) individuals control over the aspects of their identities they share publicly. This control would not be possible without technology platforms.

For Marni's LGBTQIA2 clients who are older, entry into long-term care may mean return to the closet because their gender identities are not respected, and often they experience discrimination and harassment as a result. They may have a diminished social network, and they may not be able to reside or spend time with their partners.

Technology has allowed Marni to curate herself, to express her gender, to be authentic, and feel safe "behind a screen." For example, when she discovered she had a precious short video of her infant son in the neonatal intensive care unit, recorded just before he passed away, it connected her with an emotional and pivotal time in her life. It also provided her with joy and a sense of connection to her son who she would never hold again. In the future, as she ages, she would want to be able to connect with her current son no matter the distance between them: "It's better to connect through technology than not at all."

Marni feels privileged. As a trans woman, she has experienced significant discrimination and oppression. However, because she is white, able bodied, and appears as what society says a woman should look like in a very binary world, she is afforded many privileges that so many in her community do not have. She has the autonomy to create her personal directive so that when she is not able to make decisions about her health care or living arrangements, her designated agent would be able to carry out her wishes. This is particularly relevant for her as a transgender woman because the status quo would not respect her identity and wishes based on her gender history. She worries about her clients and other members of society who identify as LGBTQIA2 and are not privileged as she is.

Technology is helpful but also has risks. While working from home during the pandemic allows her a degree of control over what aspects of her persona she wishes to share across digital platforms, she also faces challenges. For example, when communicating without visual image, she is often misgendered. While members of the LGBTQIA2 community find support on social media, there are also risks, such as doxing (sharing of personal information without permission), impersonation, and cyberbullying.

Solution. In preparation for her own aging, Marni created an advanced directive so that she can autonomously make decisions for her future care should she become dependent in some aspects of her life. She has decided that she would prefer to use technology in the future to connect with her son and social circle if they are physically separated for whatever reason, e.g., geography or another pandemic. Her services to her clients are informed by her positionality and lived experience. Hence, she advocates for a just society where there is equal access to all services, including technology, for citizens at the intersections of age, gender, race, education, religion, culture, geography, and other determinants of health and well-being. Marni believes that technology, if secure, can help individuals create a digital presence that can be used to tell their stories and foster compassion and understanding.

CHAPTER   1

What is Autonomy and Independence in the Context of Aging in an Era of Technology?

WHAT IS IN THIS CHAPTER?

In this chapter, we present definitions of autonomy and independence and how these terms are related with aging, disability, gender, and culture. We discuss how technologies can support autonomy and independence in people as they age. We provide real-world scenarios in which the meaning of autonomy and independence for people as they age are discussed.

1.1   WHAT IS AUTONOMY?

The concept of **autonomy** originated in Greek and referred to self-rule, self-determination, free will, and self-sovereignty (Dryden, 2014; Dworkin, 2015; Oxford Learner's Dictionary, 2021). Autonomy is a term used broadly to refer to liberty, self-rule or sovereignty, and freedom of the will (Dworkin, 2015). It can also equate to "dignity, integrity, individuality, independence, responsibility and self-knowledge" (Dworkin, 2015, p. 8). Autonomy is considered an individual right, and respect for autonomy is one of the ethical principles in health care (see Box 1.1). On an individual level, older adults should be respected in their abilities to make decisions concerning their own care, provided these decisions do not harm others or themselves. Respect for autonomy also applies at community levels (Schröder-Bäck et al., 2014). For example, Indigenous communities may choose to practice traditional ways of healing paired with Western medicine, referred to as "Two-Eyed Seeing" (Bartlett et al., 2012).

> **Box 1.1: Bioethics principles for older adults**
>
> 1. Justice—fairness in treatment and resource allocation, decision making that considers the other three principles
> 2. Beneficence—decision that benefits the older adult
> 3. Nonmaleficence—decision that does not harm the older adult
> 4. Autonomy—older adult has the right to make treatment decisions assuming he/she has capacity and there is informed consent

Autonomy also refers to an ability to make moral decisions guided by one's moral codes. In health, disciplines follow sets of ethical principles. Box 1.1 shows bioethics principles for services for older adults. Ethical principles do not prescribe but guide how decisions are made. For individuals providing health services, there is a balance between these principles. High-quality services are ones where patients or clients receiving the services exercise their autonomy in the decision-making process. Respect for autonomy is based on the service provider's ability to understand an older person's capacity and intersections of relevant aspects pertaining to the person, including culture, gender, race, and age, among others.

In aging, capacity to make decisions can be affected by dementia, a stroke, or other age-related health conditions. Unlike children who are required to have legal guardians to make formal decisions on their behalf, older people are adults who have a right to make their own decisions. Even when they are appointed guardians and are not able to make decisions in certain circumstances, older adults retain the capacity to make some decisions. Persons in the early stages of progressive cognitive decline prefer to participate in decision-making (Wright, 2020). Increasingly, persons diagnosed with progressive dementia are advocating for themselves and making decisions about their future care, at a time when they no longer have the capacity (see Roger's persona in Part II). Such decisions include whether and how they wish to be monitored with technologies should they be at risk of going missing. Similarly, older adults have a role in making decisions about how they are supported through technologies, such as ambient-assisted technologies (Panico et al., 2020).

In the persona featured in this Part, Marni exercises her autonomy to plan for her future care in anticipation of the challenges she will face as a trans person (Box 1.2).

Box 1.2: Autonomy and independence

"I think where I lose it—that independence or autonomy—is when people make assumptions about who I am by the sound of my voice, and then make determinations on what they're going to provide to me, or how they treat me, based on those assumptions."

"The decision to change one's body to align it with their gender identity is very personal and unique for everyone. For me, I felt I needed to make certain changes to my body and gender expression so I'm able to participate fully in society, to be treated like the person I truly am, especially if I can't ask for it and advocate for myself."

Marni Panas, March 3, 2021

The development and deployment of technologies may not always represent the needs of end users who are older. This occurs when research data does not include those from older participants and is biased toward populations represented by younger participants. Some researchers may believe that older adults with dementia cannot participate in research; as a result, the representation of older adults with dementia in research can be underrepresented. As more studies begin to include older adults and those living with chronic conditions, research on ethical principles ensures that participants provide informed consent which respects and promotes participants' autonomy, even among participants with cognitive decline.

When we consider the intersections of gender, aging, race, dis/ability, and culture, this is a perspective of "relational autonomy." According to Entwistle et al. (2010), relational thinking and accounts direct health care practitioners to attend to the impact of their interactions on clients' self-identities, life-plans, and sense of autonomy. Through relational autonomy, people may recognize potentially oppressive aspects of interactions, and develop relationships that respect and promote clients' self-governance skills.

1.2   WHAT IS INDEPENDENCE?

In Part III, we distinguish the concept of independence from autonomy, although the two are often used interchangeably. While autonomy focuses on one's ability to make decisions in all aspects of one's life, **independence** is used to describe the degree to which a person can perform a task or activity.

Health service providers assess levels of independence in self-care activities, such as bathing, grooming, and dressing, also known as basic activities of daily living. Health professionals are also interested in their clients' abilities to perform instrumental activities of daily living, or more complex activities such as banking, shopping, driving, or taking public transportation. Activities may also be related to work (employment or volunteer), leisure, spirituality, and other ways for meaningful engagement.

Independence is "idealized" in Western culture (Kirby, 2015), which "undercuts the connected nature of life" (2015, p. 17) and independence "is not an end that can be achieved and maintained" (2015, p. 17). Yet, health care, particularly rehabilitation in Western culture, assumes independence is a desirable goal for clients to strive for in interventions. This is evident in the initial, progress, and discharge assessments professionals use (see Box 1.3). These assessments typically gauge a client's progress or threshold for discharge by evaluating the extent to which the client can perform a task independently. A strong emphasis on the concept of independence conveys a value and bias toward one's ability to be self-reliant and require less help from others. This, in turn, can discourage clients from seeking help, or reinforces the reluctance of some stoic individuals from admitting to their vulnerabilities. Paradoxically, this help may make the difference that allows clients to stay in their homes in the community.

Box 1.3: Examples of assessments for older adults that examine levels of independence

- Functional Independence Measure (FIM)—uses a person's level of assistance to grade functional status from total independence to total assistance.
- Independent Living Scale (ILS)—assesses competency in instrumental activities of daily living.
- Assessment of Motor and Processing Skills (AMPS)—measures a person's performance capacity for activities of daily living and/or independent living.
- Barthel Index of Activities of Daily Living (BI)—measures a client's ability to function independently following hospital discharge.

The concept of "remaining independent" is misleading when it is used synonymously with one's ability to remain in one's home in the community. Two older adults with similar limitations in activity performance may contrast in their abilities to "remain independent" in their homes. One person may have access to resources such as home care, access to the Internet which allows the person to purchase amenities online, and socially interact with a circle of care. The other person may not be able to "remain independent" in the community because of a lack of access to these resources. Hence, an older adult's ability to remain "independent" at home is related to the level of assistance available, and not just on that person's ability to be self-reliant in activities of daily living.

1.2.1   AGING AND DISABILITY

A disability refers to a condition that is physical, sensory, cognitive, or psychological that affects a person's level of function in daily activities and is influenced by environmental and personal factors. Disabilities can be temporary, chronic, reversible, or progressive. Aging is associated with disabilities. Worldwide, about 15% of the population has at least one disability. This proportion increases to 46% among persons aged 60 years and older (United Nations, 2016). In 2018, the number of people over the age of 65 surpassed the number of children under 5 years old for the first time (Ritchie and Roser, 2019). Thus, the proportion of persons with at least one disability continues to increase with population aging. In 2018, about four in ten care recipients in Canada were seniors (Statistics Canada, 2020a). In Western culture, we associate disabilities with diminished independence. We also associate aging with disabilities (Stone, 2003). Stone advocates for a change in the social construction of dependency, and a re-conceptualization of disability. According to Stone, "we all depend on others for the provision of a myriad of services ... It would be more accurate to note that all of us are interdependent" (2003, p. 61), and "we need to re-conceptualize social life as interdependent" (2003, p. 62).

In typically developing children, the maturation process is associated with increasing independence in skills such as physical movement, communication, and self-care. While a child's autonomy also develops with maturation, certain decisions and responsibilities, such as alcohol consumption and driving, are rites of passage from youth to legal adulthood. With aging, certain disabilities can create a misperception that the autonomy of older adults is also affected. For example, while an older person may have declined sensory and cognitive processing that affects the person's ability to drive a vehicle, the individual may still have the capacity to drive under certain circumstances. The individual may also be able to make decisions autonomously, and competently, about when to drive (see Figure 1.1).

Figure 1.1: Autonomy in deciding when one's capacity to drive is optimal. Shutterstock: Photobac.

According to Stone (2003), "dependence is socially created" and "ignores the extent to which all of us are interdependent" (2003, p. 65). She further explains, "the problem is not that people are unable to do things by themselves; rather, it is that those defined as dependent are denied the right to have control over how they live their lives ... they find no contradiction between defining themselves as independent and relying on others to assist them with a variety of activities" (2003, p. 65). People with disabilities are able to live on their own with support. Nearly half (49%) of Canadians aged 15 years and older and living alone with a disability receive help with daily activities because of their condition (Statistics Canada, 2020b). Seniors also provide care to others. In 2018, almost 25% of persons aged 65 years and older cared or helped family members with long-term conditions, physical or mental disability, or challenges related to aging (Arriagada, 2020). Senior men were as likely to provide care as senior women, although the types of activities varied: men provided outdoor work and house maintenance while women provided housework, and meal preparation assistance. This type of interdependent relationship generates communal motivation and well-being in the partner who provides care to another (Le et al., 2018).

Interdependence is a concept that applies to both "human and non-human" aspects of our lives (Luoma-Halkola and Häikiö, 2020). In a study on how older adults organize their out-of-home mobility and independent living when faced with mobility restrictions, the researchers found that mobility was an "act of interdependence," and independence was "fundamentally a "collective achievement," not the result of individuals' self-reliance (2020, p. 2).

The "disability" label can create negative or positive consequences. Invisible disabilities such as mental illness or autism can create stigma with negative consequences of discrimination or bias. However, labeling can also have positive consequences by providing a diagnostic label to conditions, such as dementia. The label can create access to funding and programs or support. Indeed, people living for dementia have been advocating for "dementia" to be reframed as a disability. According to Kate Swaffer, if her condition of dementia was reframed as a disability when she was first diagnosed, she would have been eligible for benefits as an employee of a health care system, instead of being stripped of her autonomy, and sent home to wait for her cognitive function to decline (Alzheimer Society of London and Middlesex, 2021). In this case, a person in the early stages of dementia would actually be supported to live more *independently*, e.g., continue to work until the person is no longer able, while receiving employee disability benefits according to employment contracts.

For the purpose of this book, we use the term caregiver to refer to people providing care to others. The term caregiver is commonly used and understood by the general public and includes formal caregivers (i.e., paid personnel to provide care) as well as informal caregivers (i.e., relatives, friends, or volunteers) (see Figure 1.2). We are aware that other terms are being proposed, such as care partner, that reflect better the two-way nature of the care process, characterized by mutual cooperation, joint responsibilities, and opportunities to give and receive between the two parties involved in a care partnership (Olson, 2017).

Figure 1.2: Caregiver may be formal or informal, as in this case of a daughter caring for her mother. Shutterstock: Tao55.

1.2.2   GENDER

The case scenario of Marni in this Part I is an illustration of the complexities of gender intersecting with disability, autonomy, and independence, as well as socio-economics, digital literacy, and the possibility to afford technologies. Current care facilities design programs and services based on assumption of what clients need. Facilities have been designed to care for older adults as individuals with little or no consideration for a resident's relationship with a life partner. Even less consideration is given to partners of seniors who are LGBTQIA2 and who have to relocate to a care facility.

According to The Health of LGBTQIA2 Communities in Canada – Report of the Standing Committee on Health (Casey, 2019), the size of the LGBTQIA2 in Canada is difficult to estimate because Statistics Canada has not yet asked questions about gender in its surveys. The report makes 23 recommendations that address health inequities in members of gender and sexual minorities. As Canada begins to collect population data and address the Standing Committee's recommendations, we will be able to provide more equitable care for members of the LGBTQIA2 communities.

Recently, continuing care facilities have improved designs that provide a range of living options nearby, or on the same "campus," to allow partners to live within a community. This is particularly welcomed by couples who are separated because of gender, as in the case of facilities built for veterans which are almost all men. Proximity between the residents and loved ones is necessary for residents' well-being as technology cannot replace all elements of social connection. For example, an elderly man living in a veteran's facility who is hard of hearing and his spouse living in a separate facility, who is blind, would be unable to connect socially by phone or videoconferencing.

Gender disparities exist among people living with disabilities and people providing care. According to the 2017 Canadian Survey on Disability (Statistics Canada, 2018b), more women (24%) experienced a disability compared to men (20%). Earlier, it was mentioned that older men and women were equally likely to provide care. However, this is not the case among younger people. In 2018, caregivers between ages 45 and 64 showed the biggest gender disparity with women outnumbering men by about 10% (Arriagada, 2020). In a U.S. 2020 survey, 61% of caregivers of other adults were women compared to 39% men (AARP and National Alliance for Caregiving, 2020).

When caregiving is a chosen role, such as nursing, women make up over three quarters of the world average (World Health Organization, 2021), and over 90% in Western countries such as Canada (Canadian Institute for Health Information, 2020). In most countries around the world, when it comes to helping older adults living with disabilities, care providers are overrepresented by women. Data on representation of service providers who identify with LGBTQIA2 is not being collected in Canada and does not appear to exist elsewhere. Yet, groups such as the Rainbow Nursing Interest Group (https://chapters-igs.rnao.ca/interestgroup/58/about) exist to demonstrate "evidence-informed, inclusive, reflective, respectful, safe and supportive care and environments for people of all sexual orientations and gender identities and expressions" (Registered Nurses' Association of Ontario, 2020). This is an interest group of the Registered Nurses' Association of Ontario and was co-founded by Jean Clipsham and Dianne Roedding in 2007. As service providers feel safe and included with their own gender identities, the health care system will become more inclusive of patients and their families, chosen or biological.

As individuals, our sense of autonomy and the decisions we make are inextricably linked to our identities. Today's older adults arrive in the health care system with a history from the 1970's and 1980's when any gender that was non-binary was considered to be taboo. Health practice in the dominant culture still has a long way to go before marginalized older adult clients of the health care system can feel safe and valued in their interactions with service providers.

1.2.3   CULTURE

The value for individualism and independence is a minority world culture (Hammell, 2009). Yet, the narrow Western priority dominates assumptions that all older adults prize independence. In many cultures, interdependence contributes to well-being. For example, in some Latin American cultures, interdependence is understood as part of daily living in relationships among members of a social circle and drives the provision of care to a member of the family or the closer social circle who is in need (Dulcey-Ruiz, 2015). However, the Western definition of independence has reduced the scope of the concept of care in the Latin American region, and further narrows the reciprocity of interdependence to a mere contractual relationship (NU CEPAL, 2017).

A prime example in North America is the Indigenous populations. According to Indigenous ways of knowing, knowledge is interdependent with the land, its creatures (relations), the community, and time through the generations (see Box 1.4) (Battiste, 2005).

In Indigenous communities, *interdependency* and the concept of *relational* autonomy better represent Indigenous ways of knowing. Relational autonomy recognizes that people's "identities, needs, interests and—indeed autonomy—are also always shaped by their relation to others" (Dove et al., 2017, p. 150). One's influence by others may not always be ideal or in the best interest of an older person, nevertheless, respect for the person's autonomy implies respect for a person's choice to involve others in making decisions. As Indigenous peoples are inextricably tied to their land and have been historically colonized in Canada, autonomy also carries political meaning as it relates to sovereignty, self-determination, and self-governance.

Not including Indigenous people, more than one out of five people in Canada is a visible minority, or person of color. Eurocentric values that drive decisions about programs and services disregard the cultural diversity of older adults. Indeed, the concepts of interdependence and relational autonomy are integral to many cultures that call Canada home.

Box 1.4: Indigenous ways of knowing

Indigenous knowledge is passed on through observation and 'doing'. Knowledge is embedded in the natural environment and includes the knowledge and skills needed for survival. Interdependence is valued in ways including:

1. Holistic Knowledge:

- All Knowledge is connected. Indigenous knowledge is constructed as stories, traditions, skills, values, and myths that all together present a holistic picture of interdependence of humans and their environment.

2. Living Interdependence

- Indigenous people know that humans are inseparable from the land, the Earth. Traditional knowledge sees "all my relations" including all species and the Earth, which maintains sustainable, respectful, and sacred connections to the land.

3. Long-Term Time Perspective

- Circular time with a "multi-generational perspective" and a long-term sustainable viewpoint for decisions.

4. Dynamic Cultures

- Indigenous cultures have been quick to adapt to new technologies, try to improve their circumstances, modifying and adapting the colonial structures to their own purposes, while maintaining their relationship to the land.

5. Community Values

- The value of family, culture, and community is above other values. The notion of "it takes a whole community to raise a child" is much more of a reality in indigenous communities with extended families and relations. Humans are seen as part of the natural world, not the masters.

Based on Battiste, 2005; Kawagley and Barnhardt, 1999

1.2.4   TECHNOLOGY

Technologies can support autonomy and levels of independence of older adults. For example, information accessed through the Internet can inform seniors in making a decision. The Internet can also facilitate access to consumer goods and groceries. Information communication technologies allow seniors to maintain social connections and engage in online interactions. During the COVID-19 pandemic, the Internet has provided many seniors with these options, as well as access to health consultations.

In a survey conducted by AGE-WELL in July 2020, with over 2,000 Canadians 50 years and older, two-thirds (65%) of seniors, or respondents 65 years or older, owned a smartphone, compared to 59% in 2019, and 83% used it daily (AGE-WELL, 2021). The proportion of people who use video-calling doubled during the pandemic at 23% compared to the previous year. The survey also found that 88% of seniors used the Internet daily. For the respondents, technology that "manages independence" was most popular and had a positive impact on health and wellness, i.e., "techs/services such as wearable digital devices, online shopping for essential items, exercise/activity trackers and webinars/online classes."

While 76% of the AGE-WELL survey respondents aged 50 years and older stated they felt "confident using current technology" (AGE-WELL, 2021), it is unknown how representative the sample was of marginalized or racialized populations. Access to the Internet and technologies is a challenge for older adults who are of low socio-economic status, or disadvantaged by systemic racism, inequities, and discrimination. In 2017, nearly 1.4 million (42%) Canadians aged 70 years or older living in a private dwelling had a disability (Statistics Canada, 2018b). Among those living with a disability, almost 1 in 5 (19%) reported they did not use the Internet (Statistics Canada, 2020b). A "**digital divide**" refers to discrepancies in access to information communication technologies, or the Internet, between segments of our populations (Fang et al., 2018; Hilbert, 2011). People living in rural and remote locations, with lower levels of education, living in poorer communities, and people of particular racial backgrounds experience a gap in accessing technology (Choi and DiNitto, 2013b; Fang et al., 2018). Technology literacy and usability are also factors for older adults when product interface designs do not take into consideration age-related physical and sensory changes (Charness, 2020).

While AgeTech facilitates autonomy and independence among older adults, access to AgeTech is affected by social and economic factors such as education and income (Hargittai et al., 2019; Hunsaker and Hargittai, 2018). Those with low educational attainment and income are less likely to use the Internet for several reasons including the disposable income to purchase devices and an internet connection (Chang et al., 2015). Social and economic determinants of autonomy and independence also include policies for reimbursement and subsidies that make AgeTech more affordable for older adults.

Inequities may also stem from "internalized" ageism, ableism, and racism. In other words, some older adults may not feel that they are worthy of access to services or products such as technologies because they believe that they are "too old," "too disabled," or "racially inferior."

1.3   SUMMARY

In this chapter, we examined the terms "autonomy" and "independence," two words that are commonly used interchangeably to convey individualistic values of Western or Eurocentric culture. Marni's persona and scenario show us that gender identity is a factor that compels some individuals to plan for their older adulthood earlier and that technology platforms can facilitate autonomy. We also see the limitations of the two terms from the perspective of other cultures, such as Indigenous ways of knowing. An alternative perspective is relational autonomy and interdependence. In the former, decisions are made in consideration of relationships, and in the latter, there is recognition that independence and interdependence are closely related. Finally, we examined these concepts in the context of aging in an era of technology. While technology is believed to support autonomy and independence among older people, there exists the reality of a digital divide. Vulnerable, marginalized, racialized, and poor older people do not enjoy the same access to technologies as the general population. We will continue to examine these issues in future chapters.

FIND OUT MORE

1. Truth and Reconciliation Commission Reports: https://nctr.ca/records/reports/

2. San'yas Indigenous Cultural Safety Training: https://www.sanyas.ca/

3. Indigenous Canada Massive Open Online Course (MOOC): https://www.coursera.org/learn/indigenous-canada

4. SPECTRUM Aging with Pride for resources for making long-term care facilities and retirement homes safe for LGBTQ: https://ourspectrum.com/projects/aging-with-pride/

5. Centres for Learning, Research & Innovation in Long-Term Care: https://clri-ltc.ca/lgbtq/

6. Supporting Diversity and Inclusion in Long-Term Care: https://clri-ltc.ca/resource/diversity/

CHAPTER   2

International Frameworks on Health and Technology

WHAT IS IN THIS CHAPTER?

This chapter showcases different types of technologies that are intended to foster autonomy and independence in older adults and some perspectives that support the implementation and access to these technologies. The chapter defines AgeTech and the context of everyday technologies that can enhance autonomy and independence for older adults, explains the role of technologies in the enjoyment of basic human rights by older adults, and presents relevant theoretical approaches applicable to understanding the importance of access to technologies. Real-world examples are provided to understand the barriers faced by older adults accessing the technologies they need as well as the related ethical dilemmas.

2.1   TECHNOLOGY AND THE INTERNATIONAL CLASSIFICATION OF FUNCTIONING, DISABILITY, AND HEALTH

Technology is created by humans to enhance the performance of tasks or functions (Grübler, 2003). The concept of technology includes a combination of "hardware," which refers to tangible human-made things (e.g., buildings, equipment) and "software," which is the organized arrangement of human knowledge, experience, and skills (Li-Hua, 2012). Simply put, technology includes things, actions, processes, methods, and systems made by humans (Kline, 1985). The term "technology" is used to describe a range of products or services, such as high technology, or "high tech," referring to advanced technology, frontier technology, or cutting-edge technology, which are the newest or most modern machines, equipment, or methods available. These high technologies encompass various technologies developed in the Third and Fourth Industrial Revolution including Information and Communication Technologies (ICTs), robots, Internet of Things, Artificial Intelligence, smart homes and smart devices, autonomous vehicles, 3D printing and scanning, among others (Liu, 2018). In this chapter, we will discuss some specific technologies used to promote autonomy and independence of people as they age.

In order to understand how independence and autonomy can be affected by a health-related condition, we will use the International Classification of Functioning, Disability and Health

(ICF) (World Health Organization, 2002). The ICF is a classification of health and health-related domains, including the biological, the individual, and the social aspects of health. It proposes a bio-psychosocial model to understand disability. The ICF describes optimal functioning and disability as outcomes of the interaction between the health condition and contextual factors. The health condition includes the presence and severity of diseases, disorders, and injuries, while contextual factors include personal and environmental factors.

Based on the biopsychosocial model, the ICF identifies three levels of human functioning classified as functioning at the level of the body or body part, the whole person, and the whole person in a social context (see Figure 2.1). Functioning at the level of the body or body part includes **body structures and functions**. Body structures are anatomical parts of the body, for example, the brain, the eyes, structures related to the movement such as bones or muscles, or the skin. Body functions are physiological functions of body systems, including psychological functions, for example, thinking, remembering, pain, and functions of the cardiovascular system. Significant problems in body structure or function are referred to as **impairments**. The ICF presents the human functioning at the whole person level as the activity domain, which is defined as "the execution of a task or action by an individual" (World Health Organization, 2002, p. 10). Examples of activities are learning and applying knowledge, and general tasks and demands. The **participation** domain is defined as "involvement in a life situation" (World Health Organization, 2002, p. 10), in other words, participation in performing activities associated with social life roles. Examples of participation are community, social and civic life, and interpersonal relationships. Self-care activities (e.g., eating, toileting, health care management, medication management), communication, mobility, and domestic life are categories that are open to an individual or societal interpretation; thus, they can be in the domains of activities or participation. When a person experiences difficulties in performing activities, this is referred to as **activity limitations**. When a person experiences problems in involvement in life situations, it is referred to as **participation restrictions**.

Finally, at the whole person in a social context level, the ICF describes the contextual factors that influence health and functioning, including personal and environmental factors. Personal factors are internal attributes that influence how a person faces a health condition and include gender identity, age, social background, education level, lived experience, personality traits, and coping styles. Environmental factors are attributes external to the person that influence how a person experiences a health condition and include social attitudes towards personal attributes and health condition (e.g., age, gender identity, ethnicity, living with HIV, or with dementia), support and relationships, design of the built environment, products and technology, legal and social structures, and natural environment (e.g., climate, terrain), among others (World Health Organization, 2002). Notice that the products and technology are located as environmental factors in the ICF; thus, availability and access to products and technology, including services and policies, can either facilitate or hinder functioning. An example of the analysis of functioning in reference to the ICF can be the case of Michaela Schmidt below.

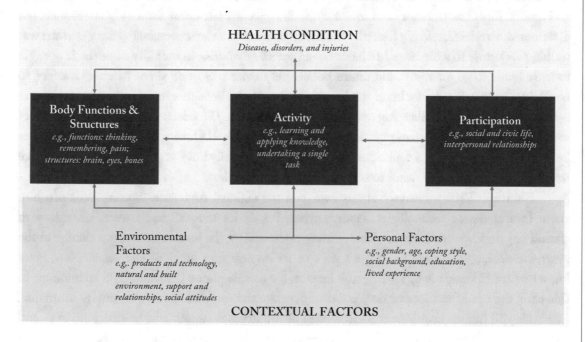

HEALTH CONDITION
Diseases, disorders, and injuries

Body Functions & Structures
e.g., functions: thinking, remembering, pain; structures: brain, eyes, bones

Activity
e.g., learning and applying knowledge, undertaking a single task

Participation
e.g., social and civic life, interpersonal relationships

Environmental Factors
e.g,. products and technology, natural and built environment, support and relationships, social attitudes

Personal Factors
e.g., gender, age, coping style, social background, education, lived experience

CONTEXTUAL FACTORS

Figure 2.1: Model of disability used in the ICF.

PERSONA, SCENARIO, AND SOLUTION

Michaela Schmidt

- **Age:** 79
- **Gender:** Woman, pronouns are she/her
- **Lives in:** Fourplex apartment in a small town in rural southern Germany with 80 year old husband Peter.
- **Social circle:** Two adult sons and their families. They live in cities 2–3 hours away by train. Dense network of neighbors and friends who live in the same town and belong to the same church.

Signature interests

Walking and cycling, volunteering with a reading program for children, attending church.

" I love being in nature. Since I was a young child, I have been going for walks and bike rides in the forest every day, rain or shine. The hardest thing about having dementia is not being free to go into the forest when I want."

Health

Dementia, mild osteoarthritis, thyroid disease.

Technology

Has a GPS device, a flip phone, and a desktop computer with Internet connection.

Persona and scenario. Michaela is an older adult who lives with dementia. She lives at home with her husband, Peter. Michaela experiences *impairment* related to structures of the nervous system and mental functions (memory and orientation). She recently left home for a walk after lunch, felt disoriented, wandered, and got lost. She was found by a neighbor who recognized her when she was walking very close to a highway. Michaela has an *activity limitation* in mobility since she is not able to leave home to go on a walk and return home safely. Peter is worried about Michaela's safety, so he decides to install extra locks on the house doors so that Michaela cannot leave home when he is occupied with the housekeeping and volunteering activities. The selection of locks as the product to manage the risk of Michaela getting lost resulted in Michaela having a *participation restriction* because she enjoys having a walk after lunch, but Peter is usually busy at that time and cannot take her for a walk. Hence, she needs to wait until Peter is available.

Solution. Peter looked for alternatives, and the regional Alzheimer Society informed him about locator device technologies. Peter purchased a device that Michaela wears on a lanyard around her neck, and it tracks her. This *new digital technology*, the locator device, is a change in the *environment* that allows her to go for a walk or for cycling after lunch as she wants. Now, Peter knows where Michaela is by looking at a map on his mobile phone that uses GPS technology. In this case, the use of the locator device did not change the impairment or the activity limitation. Michaela still has an impairment in structures of the nervous system and mental functions, and she is still unable to perform consistently community mobility on her own. However, this product addresses the restriction in her participation as she can go for a walk when she wants, while Peter can support her and knows where she is located if she gets disoriented and lost. Peter experiences peace of mind knowing that he can locate his wife when she needs assistance.

2.2   AGETECH AND EVERYDAY DIGITAL TECHNOLOGIES TO ENHANCE AUTONOMY AND INDEPENDENCE

As mentioned in the Introduction, AGE-WELL has adopted the term AgeTech which is defined as "the use of technologies, such as information and communication technologies, robotics, mobile technologies, artificial intelligence, ambient systems, and pervasive computing to drive technology-based innovation to benefit older adults" (Sixsmith et al., 2020, p. iiv). AgeTech can *enhance* autonomy and independence in healthy older adults to live the way they want. AgeTech can also *support* autonomy and independence in older adults who live with physical and cognitive impairments, and their caregivers. AGE-WELL has proposed using AgeTech because it is a more encompassing term than others, such as *gerontechnology*.

Everyday digital technologies are high tech that we use to engage in everyday activities, and to connect with people and the environment. These technologies include products such as smart devices (e.g., smartphones, smartwatches, smart homes), mobile devices (cell phones, tablets, computers), and digital communication platforms, software, and algorithms. They also include services and policies regarding the cost, literacy, and protection of personal information associated with the technology's access and use. Everyday digital technologies can empower people as they age to enjoy autonomy and independence.

AgeTech and everyday digital technologies interrelate with other technologies such as assistive products, assistive technology, environmental technology, and occupation-related technology. Through its Global Cooperation on Assistive Technology (GATE) initiative, the World Health Organization has made efforts to clarify terms associated with assistive technology. A series of publications have compiled key definitions. Table 2.1 shows key terms and definitions for understanding the relationship between some terms used currently and what is AgeTech and everyday digital technologies. As shown in Table 2.1, an assistive product is any product whose primary purpose is to maintain or improve a person's functioning and independence (Khasnabis et al., 2015). Assistive technology systems belong to health technology and comprise the application of knowledge, skills, procedures, and policies for the provision of, assessment for, and use of assistive products (Smith et al., 2018). Smith (2017) proposed the terms "environmental technology" and "occupation-related technology." Environmental technology refers to the built environment that anyone uses. Environmental technology also includes assistive products and therapeutic technology embedded in the environments, addressing different impairments, e.g., ramps or grab bars. Occupation-related technology refers to technologies people use to engage in activities or occupations. It covers a wide range of objects, tools, equipment, and methods to perform self-care, play, leisure, work, and educational activities, from very low complexity, such as a cup, to high complexity, such as a smartphone.

Box 2.1: Global Cooperation on Assistive Technology (GATE)

In response to obligations of the United Nations Convention on the Rights of Persons with Disabilities (CRPD) toward increasing access to assistive technology, WHO developed and coordinated a global initiative called the Global Cooperation on Assistive Technology (GATE). The first GATE meeting was held in Geneva in 2014. Attendees included representatives from international organizations, donor agencies, professional organizations, academia, and user groups.

The GATE goal is "to improve access to high-quality affordable assistive products globally." In order to achieve this goal, the GATE initiative is focusing on five interrelated areas called the 5Ps:

- People
- Policy
- Products
- Provision
- Personnel

Information about GATE: https://www.who.int/news-room/feature-stories/detail/global-cooperation-on-assistive-technology-(gate)

Assistive technology for older adults becomes relevant when they live with an impairment and need the assistive products to remain independent (Khasnabis et al., 2015). Assistive technology, assistive products, and environmental technologies are intended to compensate for impaired body functions or structures.

Figure 2.2: Everyday digital technologies have supported health care delivery services during the COVID-19 pandemic. Shutterstock: fizkes.

Everyday digital technologies can support autonomy when an impairment is not present. Examples include the use of a tablet and video conferencing app to communicate and socialize with family members and friends or the use of a computer or smartphone to access a website to carry out medication management-related tasks or online appointments with clinicians during a mandatory lockdown to control the spread of a virus (see Figure 2.2). Neither the hardware (tablet, computer, smartphone) nor software (apps or websites) is an assistive product; however, their use can empower older adults to execute their autonomy and keep their independence. In this example, the hardware and software act as everyday digital technologies supporting older adults to make informed decisions related to medication management and reduce social isolation resulting from a mandatory lockdown. AGE-WELL's FamliNet is a product that addresses this need. This communication tool was designed for older adult users who have impaired sensory function and allows them to connect with their social circle of friends and family.

Everyday digital technologies, environmental technology, and occupation-related technology are pervasive in our lives as we use them to engage in daily activities. Some people also use AgeTech and assistive products on a regular basis. Ideally, all technologies should be created using universal design principles. Universal design principles include: equitable use (design provides equitable access for everyone), flexibility in use (the design considers a wide range of individual abilities and preferences throughout the life cycle of potential users), simple and intuitive (regardless of the user's experience or cognitive ability, the user can easily understand how to use the technology), perceptible information (information is presented using different modes, such as visual, audible, and tactile), tolerance for error (the design minimizes the potential for unintended results and considers user's safety), low physical effort (using the technology does not require excessive physical effort), and size and space for approach and use (the design considers that objects are within environments and enough room for both is needed, e.g., design considers that people using walkers or wheelchairs need wider doors and wheelchair users or people short in size need lower water fountains) (Clarkson et al., 2013). Technologies that are not designed with universal design principles target only the average person as the user, e.g., a right-hander with a strong grip, visual acuity in normal ranges, no joint pain, technology savvy, limiting their use by older adults and other populations. When technologies are designed following principles of universal design, it is possible or easier for older adults to use them.

Technologies can be categorized by purpose. Robots that support a human user with an impairment are called assistive robots; thus, they are assistive products (Encarnação and Cook, 2017). One example is the commercially available robot Care-O-Bot 3, which can be programmed to remind older adults with memory impairment to drink water regularly, eat three meals a day, and take their medicine. In this case, a technology of high complexity (a robot) is developed to target the specific population of older adults and is intended to compensate for impairment in cognitive function. Similarly, there is an assistive robot developed by AGE-WELL researchers that prompts older adults with cognitive impairment about when and how to carry out daily activities, to support virtual visits for remote consultations with medical professionals, and to monitor the older adults' well-being (see Figure 2.3). In contrast, a robot vacuum connected to Wi-Fi, controlled remotely and scheduled through a smartphone, is not an assistive product but an everyday digital technology. This technology facilitates household tasks for older adults as well as the general population while they engage in other activities such as leisure or play.

Table 2.1: Terms and definitions on relationships between technologies

Technology and Definition	Target People Without Impairment	Target People With Impairments	Target Older adults	Address Independence	Address Autonomy
AgeTech The use of technologies, such as information and communication technologies (ICTs), robotics, mobile technologies, artificial intelligence, ambient systems, and pervasive computing to drive technology-based innovation to benefit older adults (Sixsmith et al., 2020, p. vii)	✓	✓	✓	✓	✓
Everyday digital technology High tech that we use to engage in our everyday activities and to be connected with people and our context. These technologies include products such as smart devices (e.g., smartphones, smartwatches, smart homes), mobile devices (cellphones, tablets, computers), digital communication platforms, and software and algorithms. These technologies also include services and policies regarding the cost, literacy, and protection of personal information associated with the technology's access and use.	✓	Some may be designed for those with impairments	Some may be designed for older adults	✓	✓
Assistive technology The application of organized knowledge and skills related to assistive products, including systems and services. Assistive technology is a subset of health technology (Smith et al., 2018, p. 474).		✓	Only those with impairments	✓	Only when an impairment is present
Assistive product "Any product (including devices, equipment, instruments, and software), either specially designed and produced or generally available, whose primary purpose is to maintain or improve an individual's functioning and independence and thereby promote their wellbeing" (Khasnabis et al., 2015, p. 2229).		✓	Only those with impairments	✓	Only when an impairment is present
Occupation-related technology "The technology used by everyone in their everyday activities, such as the telephone, the computer, the bicycle and the television." (Smith et al., 2018, p. 474).	✓			✓	✓

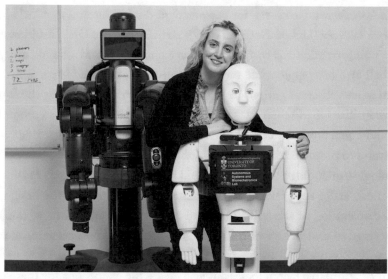

Figure 2.3: Goldie Nejat (University of Toronto) with Casper, a socially assistive robot. Photo by John Hryniuk. Courtesy of AGE-WELL.

2.3   TECHNOLOGIES FOR THE ENJOYMENT OF BASIC HUMAN RIGHTS IN AGING

It is widely recognized that assistive technologies are critical mediators for people with disabilities to exercise their human rights and fundamental freedoms, enhancing the possibilities to participate in and control their environments (Borg et al., 2011; Desmond et al., 2018; Khasnabis et al., 2015). The adoption of principles of a human rights framework by global organizations, social movements of people with disabilities, academics, and researchers, has set that everyone who needs assistive products has the right to access them, so the individual health and well-being needs are met (Desmond et al., 2018). Access to assistive products empowers users to enjoy other human rights such as work, education, rest, and leisure, move around freely, receive information, and participate in a community's cultural activities. Access to technology also supports other human rights such as proper health care and social services; thus, nations have an obligation to create strategies to ensure that all human beings benefit from scientific and technological advancement. Unmet needs due to inaccessible technology lead to inequalities in the enjoyment of human rights.

The United Nations Convention on the Rights of Persons with Disabilities (CRPD) was declared on 2006 with a view of persons with disabilities as "subjects with rights, who are capable of claiming those rights and making decisions for their lives based on their free and informed consent as well as being active members of society" (United Nations, 2007). To date, the CRPD was ratified by 182 states; Canada formally ratified it on March 11, 2010. The CRPD recognizes that individual *autonomy* and *independence* are essential for persons with disabilities and this includes the freedom to make their own choices (United Nations, 2007). Article 9 of the CRPD addresses accessibility, including access to information and communications technologies and systems, and points out that States Parties shall also take appropriate measures "to promote access for persons with disabilities to new information and communications technologies and systems, including the Internet" (United Nations, 2007, p. 9). The CRPD then recognizes that to fulfill its purpose "to promote, protect and ensure the full and equal enjoyment of all human rights and fundamental freedoms by all persons with disabilities, and to promote respect for their inherent dignity" (United Nations, 2007, p. 4), the access to technologies, including the Internet should be guaranteed by States Parties. Article 9 also declares that States Parties shall take appropriate measures "to promote the design, development, production and distribution of accessible information and communications technologies and systems at an early stage so that these technologies and systems become accessible at minimum cost" (United Nations, 2007, p. 9). Thus, nations are obligated to develop strategies to promote not only the development of affordable technologies but also create the conditions for these technologies to reach a commercialization state, making them available to the individuals who need them.

Based on the CRPD, access to assistive technology is a human right and the CRPD creates an international, legal obligation for states to develop and implement effective public assistive technology provision systems (Borg et al., 2011; de Witte et al., 2018; Khasnabis et al., 2015). The provision of assistive technology is usually part of the national or regional health care and welfare systems (de Witte et al., 2018), which has improved access to the assistive products labeled as assistive technology (or the "assistive technology list") in each country or region. However, some technologies, such as augmentative and alternative communication systems, technologies for visual impairment, and cognitive augmentation are not on the list of some countries. This results in people with hidden impairments (e.g., limitations in vision, speech, cognition, or memory) having little or no opportunity to access the assistive products they need (Rios et al., 2014). In 2016, the World Health Organization developed the Priority Assistive Products List to improve access to high-quality, affordable, assistive products in all countries. This list includes 50 priority assistive products, which were selected on the basis of widespread need and impact on a person's life and were stated as "an absolute necessity to maintain or improve an individual's functioning" (World Health Organization, 2016, p. 1). In addition to the traditional assistive products (e.g., wheelchair or communication boards), the list includes products such as fall detectors, simplified mobile phones, GPS devices, personal digital assistant, and video communication devices. The inclusion of these new technologies has the potential to improve access to everyday digital technology to enhance independence and autonomy in people with disabilities.

The CRPD states that people with disabilities must access new information and communication technologies and systems, including the Internet. However, the right to access these everyday digital technologies that are not classified as assistive technology but support the individuals' autonomy and independence has not been addressed to the same extent as the assistive technologies. Examples are products such as smartphones, computers, smart home systems, and the Internet. Lack of access to these everyday digital technologies creates inequities. These inequities contribute to the digital divide between those who are able to benefit from the existing digital technologies and those who are not. The digital divide does not refer merely to the lack of access to digital technologies by some social groups but to the ineffective usage and lack of real impact in users' lives. Elimination of the digital divide implies that all members of a society can fully adopt the technologies they need. The digital divide has been explained using a social network framework where the technology adoption (also called diffusion of technology) is influenced by the nature of the ties among individuals (the network structure) and by the characteristics of each individual (the personal adoption threshold). Technology adoption takes a certain amount of time. A divide occurs as a result of the prolonged time for innovations to spread through social networks that have particular characteristics that put them at a disadvantage (Hilbert, 2011). The digital divide between industrialized and developing countries has noticeably increased since 2002. Despite the similar number of technology devices among nations, the information capacity (i.e., the capacity to communicate through mobile devices and the Internet in optimally compressed kilobits per second per capita) is not. Individuals living in industrialized countries have had 15 times more bandwidth available than those in developing countries in 2007, and the trend has been rising (Hilbert, 2011). Thus, geoeconomic factors count for the digital divide, but they are not the only ones.

Other factors that contribute to the digital divide in middle-aged and older adults include sociodemographic variables such as education, income, and age. People with higher education and income levels tend to have more access to ICT, while people older than 70 years of age have less access to ICT than younger adults. Among the factors are little or lack of skills or familiarization with digital technologies, poor access to an internet connection, costly access to internet plans that are not affordable, and design of the products that do not incorporate universal design principles, making it difficult for older adults to use them (Fang et al., 2018; Kottorp et al., 2016). When older adults cannot use everyday digital technology to access health care services that are increasingly reliant on ICTs for delivery (e.g., telehealth or scheduling appointments using digital platforms), older adults are marginalized. Older adult populations' rights to health promotion and well-being are affected (Kottorp et al., 2016). This is evident during the COVID-19 vaccination rollout, where the limited access to quality internet, and poor usability of web-based platforms have hampered seniors' ability to book appointments for vaccines. The persona and scenario of Gladys help to exemplify the digital divide in older adults.

PERSONA, SCENARIO, AND SOLUTION

Gladys Rincón Neira

- **Age:** 70
- **Gender:** Woman, pronouns are she/her
- **Lives in:** Single family condo alone, in Bogotá, D.C., Colombia
- **Social circle:** Family and a few close friends

Signature interests

Sewing, reading, yoga, religious activities, volunteering, spending time with her granddaughter. She is the main caregiver of her mom, who is 91 years old.

" I feel able to do things as I did before, in the old way, since the pandemic began, many things have to be done online, I feel as if I have lost the ability to do them, and I don't like that because I feel like I'm not myself. "

Health

High blood pressure, osteoarthritis, and fibromyalgia.

Technology

Uses a smartphone and a laptop computer. Has no Internet connection at home. Uses smartphone data to keep in contact with her family.

Persona. Gladys is a 70-year-old woman born in Colombia. She was married and had two daughters; one lives in Colombia nearby, and the other lives abroad. Gladys divorced when she was 33 years old and never married again. She lives alone in a condo and has no Internet connection because she feels she does not need it. Gladys has chronic medical conditions including high blood pressure, osteoarthritis, and fibromyalgia, although she is independent in all her basic and instrumental daily activities. She drives and visits her 91-year-old mom regularly, who lives with dementia in a long-term care facility. Gladys is her mother's substitute decision maker and has been her caregiver for many years. She manages her mother's finances and health care services. Gladys enjoys doing yoga, sewing, reading, and attending church. She is also involved in volunteering with relatives and friends, such as providing rides to older adults to attend medical appointments. She has a 2-year-old granddaughter with whom she loves spending time and caring for when her daughter needs support.

Scenario. Many of Glady's interests and life roles have been impacted by the measures to control the spread of the COVID-19 pandemic. Gladys used to be independent in all her daily activities, including health and financial management for her and her mother. She also used to be heavily involved in activities with her family, friends, and community. Gladys also used to support her daughter, who lives abroad, in managing her daughters' properties. For example, she used to attend the condos' annual general meetings (AGM), pay utilities bills and taxes for her daughter's and her properties. All these activities were done in person. She did not feel she needed an internet connection at home. When the COVID-19 pandemic hit the city, everything changed for Gladys. Now, all the activities she used to do independently in person were moved online. She found the internet platforms (e.g., for e-banking or health management) hard to understand

and use. She also is afraid to use them due to the risk of being scammed. She feels she needs an internet connection now, but she does not want to increase the risk of getting COVID-19 by giving the technical personnel access to her condo to install the internet. Now, she relies on her daughters to manage the bank account, health services, bill payments, or attend the AGM as all of them are now online. She is also sad because she cannot spend time with her granddaughter. The current situation makes her feel isolated from family and friends and frustrated as she cannot carry on many of the daily activities she used to do independently or engage in meaningful life roles.

Solution. Gladys' situation is similar to many other older adults worldwide who live in countries or regions with limited internet connections and do not feel confident using digital platforms to manage personal information. During the COVID-19 pandemic, many older adults have not had enough time to confidently transition from in-person to online to carry on their daily activities. The pandemic suddenly hit their environments, and they found themselves having little skill or familiarity with digital technologies and no access to an internet connection. Providing free internet connection and educational support to increase older adults' knowledge and skills in using online platforms would prepare them for future pandemic scenarios so they will not lose independence when the use of technology becomes a necessity.

2.4   SUMMARY

Younger generations of adults who are preparing themselves for their future older selves, such as Marni, may be more familiar with everyday digital technologies than older generations. However, barriers to internet access in terms of signal and affordability may persist when Marni becomes an older adult if changes are not put in place now. These changes should consider the intersection of factors that contribute to the digital divide: race, age, education, culture, and gender, among others. Conscious or unconscious biases create barriers to technologies for some marginalized social groups, negatively affecting their personal autonomy, and the experience of equal opportunities for social participation. An effective change to ensure equal access to the technologies needs *government leadership in policies* and regulations with a commitment to the necessity of everyday digital technologies in the current world, *corporation and business practice change* to make necessary services affordable, and *education change* that promotes the citizens to address conscious and unconscious biases in the society, and the implementation of education programs to increase technology literacy among all citizens, including older adults.

FIND OUT MORE

1. WHO Priority Assistive Products List: https://www.who.int/news-room/feature-stories/detail/priority-assistive-products-list-(apl)

2. GATE: https://www.who.int/news-room/feature-stories/detail/global-cooperation-on-assistive-technology-(gate)

PART II

How Can Technology Support One's Autonomy?

"We want autonomy for ourselves and safety for those we love. That remains the main problem and paradox for the frail. Many of the things that we want for those we care about are things that we would adamantly oppose for ourselves because they would infringe upon our sense of self."

– Atul Gawande

PII.1 THE CHALLENGE

Autonomy is something that we all value. Being autonomous is about being able to have control over our decisions, identity, and activities that are meaningful to us. It is the ability to choose which makes us human (L'Engle, 2001) and provides us with the greatest sense of quality of life, satisfaction, and overall well-being (Kroemeke, 2015). Unfortunately, as one ages, our ability to have autonomy over our lives becomes threatened. Aging for many individuals comes with various illnesses and functional limitations that eventually lead to difficulties in performing activities of daily living (Canadian Institute for Health Information, 2011). When ill, we often experience greater dependency, and with it, a greater likelihood that our interests and values will be overridden by others (Sherwin and Winsby, 2011). Clinicians for example are at a significant risk of paternalism where they are tempted to substitute their own judgment for that of the person they are providing care for, especially when they feel that they have better insight into the person's needs than the person receiving care and in order to ensure the best medical outcome (Sherwin and Winsby, 2011).

The same risk can be said among family caregivers, where they must become responsible for the life choices of the person they are caring for when they are no longer able to make such decisions on their own (Smebye et al., 2016). In many circumstances, these decisions are triggered reactively following a recent diagnosis of an illness or hospitalization, forcing them to make hurried decisions without enough time to find out their options. In others, it is more of a steady process with the caregiver slowly taking on increasing roles and responsibilities as the health of the person they care for declines (Adekpedjou et al., 2018). Do these decisions reflect what the person would have wanted? Or, as noted by Atul Gawande, are caregivers, both paid and unpaid, placed into a situation where their values are in conflict, where these choices infringe upon their own sense of self? Are there existing technologies, models, or theories that could be applied to ensure an older adult's sense of self and self-determination are respected while still ensuring that they remain safe?

PII.2 WHAT IS IN PART II?

In this part of the book, we consider how present technologies can support autonomy, such as self-identity, capacity, and advance care planning. During the last half of this part, factors such as risk and privacy will also be explored in terms of their intersection with technology use, and how it can inevitably influence the autonomy of older adults and persons living with dementia. Therefore, this part will be broken down into five chapters as follows:

- Sense of self and identity

- Capacity

- Advance Care Planning

- Risk

- Privacy

Each of these chapters begins with a description of the underlying topic and its role on the autonomy of older adults and persons living with dementia. We then showcase examples of some of the technologies related to these topics that can assist in this population's sense of autonomy. Finally, ethical tensions and their influence on autonomy will be woven in. In these chapters, we also weave in the real-life stories featured in our personas and scenarios.

PERSONA, SCENARIO, AND SOLUTION

Roger Marple

- **Age:** 63
- **Gender:** Man, pronouns are he/him
- **Lives in:** Apartment in Medicine Hat, Alberta with his cat, Bernie
- **Social circle:** Aside from his cat, he has three adult children who live in British Columbia and Alberta and two grandsons. He also has a brother who lives in Ontario and keeps a lookout for him remotely. Roger is surrounded by a close circle of friends that span across multiple provinces in Canada

Signature interests

National and international dementia advocate, golf, biking, traveling, community events, baking, and spending time with his cat.

"Being proactive is everything. Using proactive strategies like Life360 and Google Home allows me to continue to live well with dementia and make the decisions that are important to me, such as living alone."

Health

Progressive dementia.

Technology

Has a computer, smartphone, and an Internet connection. Uses Life360 and Google Home.

Persona. Roger is 63 years old, single, and lives alone in a four-story apartment building in Medicine Hat, Alberta. He is the proud father of three grown children and has two grandsons. He enjoys golf, mountain biking, travel, community events, knows his way around the kitchen with a real appetite for baking, and enjoys spending his downtime with his beloved cat, Bernie the "wonder cat." Roger worked for Alberta Health Services and has worked in supply management in the south zone for over 23 years. He also lives with progressive dementia. Roger was invited to join the Alzheimer Society of Canada's advisory group to help raise awareness of the needs of people with dementia, including the specific needs of people living with young onset and/or early-stage dementia. Roger also serves on the board of directors of the Alzheimer's Society of Alberta and Northwest Territories and is active in supporting dementia research in Canada. Since his diagnosis in 2015, Roger has made it his mission to dispel myths about the disease and the stigma associated with dementia. He is a firm believer that you can live well with this disease regardless of challenges and is passionate about sharing his message of hope.

Scenario. Even though Roger remains independent and active in his community, he worries about his future. He knows that his dementia journey will eventually lead to him making tough decisions, such as when he will need to enact his Power of Attorney, and who will be included in his circle that will be responsible for making decisions for him when his dementia no longer allows him to. He lives alone in Medicine Hat, with most of his family spread across Canada, which further complicates things. Being able to be his own decision maker, he has been able to continue to engage in activities his family sees as being "risky," such as going for solo drives to the mountains or going for midnight strolls at his favorite park. What will his life look like when his doctor and family start to make his decisions for him? Will they ask for his consent even when society no longer deems it as necessary? Or will decisions be made without consulting with him first? At the same time, he fears that he will soon lose his sense of self and identity; that people will only see him for his dementia diagnosis, and not for the incredible life he has lived up until this point.

Solution. While Roger's situation is complicated, he is fortunate to have a very supportive family doctor who has been an incredible resource in terms of identifying strategies, including technologies, to help him maintain his autonomy and independence for as long as possible. Within the context of autonomy, there is promise of several existing and emerging solutions that could be used to assist him. As will be discussed in the next five chapters, digital storytelling could assist him in continuing his legacy of all of the advocacy work he has done up until this point to combat the stigma associated with dementia. Self-sovereign identity and guardianship could serve as a means of providing him with a greater sense of autonomy over who owns and controls his personal information; and underlying concepts, such as the Goldilocks Principle on Dementia and Wayfinding could enable Roger to adopt proactive strategies so he can continue to engage in occupations that are most meaningful to him, such as going on road trips and living on his own in Medicine Hat.

CHAPTER 3

Sense of Self and Identity

WHAT IS IN THIS CHAPTER?

In this chapter, we define sense of self and identity, and examine how they change during the aging process. Then, we describe the role of a sense of self and identity on autonomy and how factors such as culture, social environment, and the presence of impairments influence them. Then we identify the role of technologies such as digital storytelling on identity. Throughout the chapter, we also highlight the tensions between a sense of self, identity, and autonomy in the aging process.

3.1   WHAT IS A SENSE OF SELF AND IDENTITY?

Simply put, identity refers to "the individual's definition of who one is" (Kohon and Carder, 2014, p. 48). Identity is not a pure internal construction of self as an isolated identity as it relates to our relationship with society. Identity is linked to a sense of belonging and appreciation by others. Identity forms an organizing cognitive–affective schema through which individuals interpret their experiences and that, in turn, can be altered by experiences (Dollard et al., 2012).

When we construct our sense of self, we use different sources of information. Neisser's theory of self-knowledge (2008) proposes five different sources of information that we use to develop our sense of self: the ecological self, the interpersonal self, the extended self, the private self, and the conceptual self (see Figure 3.1). The **ecological self** refers to how people perceive themselves in a relationship with the physical environment in terms of place and activities performed. The **interpersonal self** refers to how people perceive themselves in a relationship with the social environment regarding emotional rapport and communication exchange with other human beings. The **extended self** is built from personal memories, experiences, and routines people engage in regularly. The **private self** is drawn from the unique experiences that the person realizes are not directly shared with other people. Finally, the **conceptual self** or "self-concept" is an abstract representation of the self and is developed from experiences in social roles (e.g., I am a wife, a son, or a teacher), beliefs about what a human is (including spirituality) and does, and trait attributes, for example, I believe that I am strong, I believe that I am healthy, or I believe that I am clever. We define ourselves based on different information sources, both internal and external. This process may create contradictory notions of us that we need to reconcile.

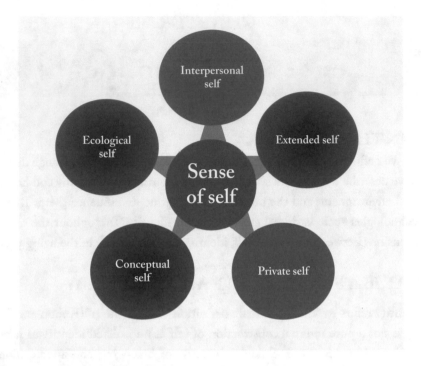

Figure 3.1: Sources of information to develop our sense of self.

The aging process can create contradictions in our sense of self due to changes in capacities, abilities, roles, and routines we face during this process. For example, a person can have the representation of being a good cook, a cognitive decline can affect one's capacity to remember the recipes, thereby affecting their ability to prepare the desired dishes. This person's identity as a good cook is then challenged by the final result of a dish that does not meet one's own expectations if they have forgotten an important ingredient or step in the cooking process. The contradiction occurs when the abstract representation of the self and the perceived performance or output in the attempt to do an activity do not match. The changes in the sense of identity have been reported in older adults and seem to occur at the same rate in older adults at early stages with dementia and as in healthy older adults (Caddell and Clare, 2013).

The changes in the sense of identity during the aging process can be explained by the intersection between cognitive and psychological processes. The Identity Process Theory explains how adults use three cognitive processes in negotiating new experiences associated with the aging process and their psychological sense of self throughout adulthood. The first process is "Identity Assimilation," in which adults experience physical, cognitive, and social decline and try to maintain a sense of self that was developed on previously established cognitive and affective representations (schemas) about the self. Through the second process, "Identity Accommodation," adults make changes in their identity in response to the new experiences that conflict with existing self-schemas. In the third process, "Identity Balance," adults adjust their identity to incorporate new age-related relevant experiences while maintaining a consistent representation of the self. Identity balance allows adults to reflect and evaluate themselves realistically as they age while preserving the essence of their sense of self (Westerhof et al., 2012). Adults who use predominantly identity assimilation or accommodation may have a more difficult physical aging experience than those who employ identity balanced to face the changes of aging. The assimilators may continue doing activities for which they no longer have the physical capacity, while the accommodators may overreact to signs of aging, restricting themselves from engaging in activities they used to do (Whitbourne et al., 2002).

3.2   ROLE OF SENSE OF SELF AND IDENTITY ON AUTONOMY

Because identity is developed based on internal and external sources of information, how adults use the processes to adjust their identity as they age depends on factors such as the culture, social environment, and presence of impairments. One critical factor for developing and maintaining the identity is the culture into which adults are immersed and how this culture values old age and youth. For example, in a youth-oriented culture, such as the Western culture in which representations of youth are valued at the expense of old age, adults can use identity assimilation to protect them from the harmful consequences of perceiving themselves as getting older and losing value in their society. In contrast, in the same culture, identity accommodation can lead individuals to have a more negative experience of the aging process (Westerhof et al., 2012). In other cultures, in which old age is more valued, identity accommodation does not necessarily lead adults to have a negative experience of their aging process.

How family, caregivers, and health practitioners perceive aging can also impact the identity of an older adult. Some cultures may have strong ageism beliefs, e.g., aging is an age of decline and withdrawal from public and social spaces (Wilińska et al., 2018). Such beliefs will impact the way in which older adults are treated and the services they receive. In cases like this, the interpersonal self, one of the sources individuals use for developing the identity, is fed by ageist behaviors from members of a society who are not yet older adults. An older adult whose self-concept includes "I am strong" and "I am active in my community" may be treated by protective family and caregivers as a fragile person who should not engage in community activities because of concerns of the risk of falling. This older adult needs to deal with the conflict between his or her positive self-concept and the negative interpersonal self from the interactions with others.

An impairment can also impact the identity of older adults. For example, in the case of those living with dementia, the progression of dementia affects different aspects of identity related to the extended self and self-concept, including role identities and self-knowledge of personality traits. Also, older adults living with dementia may have an outdated version of their identity (Caddell and

Clare, 2013). This outdated identity can result from the cognitive impairment that may prevent older adults living with dementia from using the cognitive processes of assimilation, accommodation, and balance to negotiate new experiences of the aging process while preserving the sense of self. Another example is the significance that falling has for an older adult's identity. It has been shown that older adults perceive likelihood of falling as a "threat to identity." They reported using strategies to protect themselves against being seen as someone at risk of falling which ensured the maintenance of an identity defined by being physically competent (Dollard et al., 2012).

For older adults with dementia, the cultural and social environments they are immersed in have a critical role in preserving the sense of self and identity. The literature describes several processes and interactions that depersonalize a person with dementia, including disempowerment, labeling, infantilism, and objectification. Labels and words with negative connotations to refer to the person with dementia depersonalize individuals and may lead some to be treated in ways in which their personhood is not supported by caregivers. Depersonalization diminishes the autonomy and personal choice of older adults with dementia and other conditions (Fazio et al., 2018).

A cultural change in which caregivers have shifted from a biomedical care approach to a person-centered care approach has demonstrated beneficial effects on the quality of life of older adults with dementia, including improvements in perceived individuality, autonomy, dignity, and security (Li and Porock, 2014). Thus, there are several ways in which family, caregivers, and health care providers can uphold the identity of older adults with dementia and, in turn, benefit the sense of autonomy and quality of life. Small and colleagues propose some strategies as follows (Small et al., 1998).

- Caregivers can reinforce the self of residents by addressing the older adult by their name. Some older adults may prefer being addressed by their first name, while others prefer being addressed by title. Caregivers need to respect a person's autonomy by finding out more about each resident's preferred form of address and making an effort to accommodate them.

- Caregivers should acknowledge older adults' verbal and nonverbal presentation of self in conflicts to achieve a mutually satisfactory resolution.

- Caregivers can help older adults preserve the sense of self and identity by collaborating in the co-construction of the residents' preferred personae (i.e., the public aspect or roles older adults want to show in society). It will require exploring the residents' background and acknowledging the residents' interpretation of their needs, desires, and choices for certain forms of care.

- Staff should be sensitized to respect the autonomy of residents who may prefer different levels of independent behaviors while at the same time meeting the residents' care needs.

Box 3.1: Autonomy and self-creation

"Recognizing an individual right of autonomy makes self-creation possible. It allows each of us to be responsible for shaping our lives according to our coherent or incoherent—but, in any case, distinctive—personality. It allows us to lead our own lives rather than be led along them, so that each of us can be, to the extent a scheme or rights can make this possible, what we have made of ourselves. We allow someone to choose death over radical amputation or a blood transfusion, if that is his informed wish, because we acknowledge his right to a life structured by his own values" (Dworkin, 1993, p. 224).

3.3   TECHNOLOGIES TO SUPPORT SENSE OF SELF AND IDENTITY

Technologies can support older adults' sense of self and identity indirectly or directly. Indirectly, by supporting the independence or autonomy of individuals with chronic conditions whose self-concept is being an independent person (e.g., through use of home monitoring technologies). Other technologies can directly support the preservation of the identity in older adults with dementia, for example, digital storytelling.

Home monitoring technologies, including telecare systems using ICT, can support the independence and autonomy of older adults for disease and health management. When older adults feel more independent, their identities can be supported as well. Research shows that older adults who have telecare services using ICT think that the system permits them to retain their independence and enjoy positive social participation, which supports a positive sense of identity. Older adults also value that telecare allows them to stay at home when requiring specialized care as their homes and possessions are important elements of their identities (Bowes and McColgan, 2013).

Technologies can also support a sense of self and identity more directly when used to support memories and uphold the personhood of an older adult. These technologies include those assistive technologies that compensate for memory impairments and technology interventions using ICT to create digital stories that preserve and share relevant elements of the identity of an older adult.

Tensions arise when older adults perceive that using technologies that support independence threatens their identity. Studies suggest that older adults accept technologies only if they perceive a need to use them, and the technologies do not damage their sense of self and identity (Mccreadie and Tinker, 2005). Also, older adults can resist the use of AgeTech such as assistive robots or wireless monitoring systems because they are perceived as reinforcing negative stereotypes of aging such as "being frail," "being dependent," "being inactive," or "being incompetent" (Wu et al., 2014). Older adults may also feel that technologies, such as robots, will take away or reduce human contact and care (Fristedt et al., 2021). In older adults, the acceptance of AgeTech is influenced by negotiating the new experiences faced during the aging process and their psychological sense of self (McGrath et al., 2019). Thus, when AgeTech or assistive technologies do not support older adults' identities, users will either not accept or abandon the devices.

3.3.1   ASSISTIVE TECHNOLOGIES TO SUPPORT MEMORY IMPAIRMENTS

Memory impairments impact the sense of self and identity. Older adults with memory impairment have difficulty remembering events that happen either recently or in the far past, thus affecting the representation of the self. Most of what is currently known about the potential benefits of these technologies comes from research studies that have used technologies to curate and capture life events to support older adults' recall and share their episodic memories, thus supporting reminiscence. However, little research has investigated how these technologies can strengthen the sense of self in older adults (Sas, 2018).

Figure 3.2: SenseCam. Photo by Rama, CC BY-SA 2.0 FR via Wikimedia Commons.

Examples of technologies to curate and capture life events are Memento, a system that was developed to support the creation of scrapbooks in both digital and physical form, supporting older adults to reminisce (West et al., 2007). Another example is SenseCam shown in Figure 3.2 (Hodges et al., 2011). This wearable device automatically captures photos of recent daily life events that can then be presented to older adults with memory impairments to support remembering (Hodges et al., 2011). These technologies are being called personal memory technologies and are targeting not only users with memory impairments but also the general population (Crete-Nishihata et al., 2012).

3.3.2   TECHNOLOGIES TO SUPPORT IDENTITY: DIGITAL STORYTELLING

Information and communication technologies have been used to facilitate storytelling in older adults. A digital story is a short clip that wove together sequences of still images and photos, narration, music, video clips, or written text, used to share messages to different audiences, including family, friends, care providers, community members, and the general public (Stenhouse et al., 2013; Waycott et al., 2017). The findings of a recent systematic literature review show that digital storytelling is used primarily with older adults with mild cognitive impairment or dementia and less commonly with older adults with cancer, visual impairment, Parkinson's disease, diabetes, depression, and other chronic conditions (Ríos-Rincón et al., 2021). The utilization of digital storytelling has two primary purposes: (1) as a therapeutic means to support memory, reminiscence, identity, or self-confidence in older adults living with mild cognitive impairment or dementia; and (2) to facilitate conversation and social connection between persons living with dementia and their families (Ríos-Rincón et al., 2021).

Digital storytelling can support identity in older adults with dementia during the process of creating the stories because it stimulates enjoyable memories for the older adults with dementia (Massimi et al., 2008). Creation of a digital story is also seen as an art-based activity, a creative process that can support the identities of older adults living with dementia (Stenhouse et al., 2013). Also, during the sharing stage, in which usually older adults share the final product, the short video clip, with a selected audience (e.g., the older adults' families and caregivers) promotes satisfaction. The stories can also encourage conversation about past events and help the staff in long-term care facilities to know more about the lives and identities of the older adult residents. Thus, digital storytelling can help older adults with retrieving, organizing, and using memories with the purpose of creating and sharing the stories. Digital storytelling can also help younger family members know more about who the loved one is. In health care settings, digital storytelling can help staff learn more about the lives and identities of the residents with dementia, which can reduce conscious or unconscious bias against aging, allowing them to provide person-centered care with respect and dignity.

The facilitators of digital storytelling play a critical role in enabling the autonomy of older adults with dementia. During the process of creating the digital stories, the facilitators use key communication strategies such as active listening, strategic questioning, comfort with silence, and therapeutic responding to ensure collaboration with the older adults to select the topic and create the stories. These intentional interactions where facilitators act as weavers, pulling narrative threads together to create the story, can strengthen an older adult's sense of self and identity (Hollinda et al., 2019).

CHAPTER 4

Capacity

WHAT IS IN THIS CHAPTER?

This chapter introduces, describes, and discusses the role of capacity in autonomy among older adults and persons living with a cognitive impairment. We identify the role of potential solutions such as the integration of specialized mobile interfaces on autonomy among this population. Finally, we highlight the tensions between consent and capacity, and its implications on autonomy for this population.

4.1   ROLE OF CAPACITY IN AUTONOMY

Individual autonomy embodies the mainstream understanding of autonomy where one has the capacity and freedom to make one's own choice, as long as one understands the choice being made (Department for Constitutional Affairs, 2007). Within individual autonomy, as discussed in the previous chapter, a person is able to conduct one's life in a manner of one's own choosing where it is seen as being one's own person, directed by conditions, desires, or characteristics that are part of what can be considered a person's authentic self (Christman and Anderson, 2005). **Relational autonomy** (Nedelsky, 1989) is an alternative approach to autonomy which requires that attention is paid to the ways that people exist within relations of social support and community. As highlighted by Harding (2012), neither version of autonomy includes a right to having one's wishes carried out, but rather the right to make the relevant decision. Indeed, this view of individual autonomy is appealing, in part because people generally want to think of themselves as having the ability to choose their own path in life, to make decisions free from constraints, and to decide which values they base their life on.

Autonomy is challenged when an individual is considered to no longer have the **capacity** to make informed decisions. The concept of capacity is one of the foundational elements that define decision making (Law Commission of Ontario, 2020). When we are concerned with issues such as consent, capacity can be seen when the person has the ability to weigh the advantages and disadvantages of a particular course of action and use higher levels of reasoning to come to an informed decision (Naffine, 2005). Examples can include decisions about whether to sell one's house, or whether to undergo a major surgery that was proposed by a doctor. Capacity also refers to an ability to understand the difference between right and wrong and make a moral decision that moves away from what is wrong. For example, the judgement to understand that stealing is wrong and that one needs to pay for groceries before leaving the store. Under existing legal frameworks, capacity determinations are decision-specific, subjective, and concerned with an individual's ability to make a particular decision at a specific time (Naffine, 2005).

Receiving a diagnosis of a form of cognitive impairment poses challenges to the "desired view" of autonomy, where one is able to freely make decisions based on one's values. This is because key symptoms of dementia can include a diminished ability to rationalize one's actions and decisions. This can be seen in the difficulty to grasp new ideas or being unable to effectively communicate one's thoughts (Alzheimer Society of Canada, 2021). Judgments of an individual's capacity are made informally by health care professionals and family members and can be relatively straightforward, particularly when the person is incoherent when engaging in conversation, retains little information, and shows little insight into the consequences of a decision (Alzheimer's Association, n.d.). Capacity assessments are typically conducted as a team effort, led by physicians and psychologists. If they meet training requirements, other health care professionals, such as nurses, social workers, and occupational therapists can also be designated as capacity assessors (Government of Alberta, 2021a). Capacity assessments within Canada include a mixture of interviews, observations, and standardized assessments such as the Mini-Mental State Examination (Folstein et al., 1975) and Montreal Cognitive Assessment (Nasreddine et al., 2005). This information is used to assist the assessor with their overall recommendations (Seniors First BC, 2021).

Capacity assessments may not be definitive due to the fluctuating levels of lucidity that is often seen among those living with dementia that allow for significant decision making (Alzheimer's Association, n.d.). When an individual is first diagnosed with dementia, for example, they are most often fully able to make coherent decisions on their own. As their disease progresses, the times when the individual is lucid become more and more unpredictable, making the timing for these assessments difficult to determine. It is also important to note that the decisional involvement of people living with dementia may not always be attributable to the factors associated with their disease.

4.2   SOLUTIONS TO SUPPORT CAPACITY

As highlighted in this chapter, the concept of capacity in relation to an older adult's autonomy is complex and results in an eventual loss of an individual's ability to make important decisions, especially among persons living with a cognitive impairment. While this creates worry and concern for many, such as Roger and Michaela who are featured in the personas, there has been an increase in the number of strategies to help persons living with dementia and their caregivers. While this part of the chapter does not encompass all tools and strategies that are available in the area of capacity, it provides the reader with examples of emerging solutions.

4.2.1 ALTERNATIVES TO CURRENT STANDARDIZED TESTS TO MEASURE CAPACITY

The cognitive status of older adults is commonly assessed during visits to health care settings by health care practitioners trained in the administration of standardized cognitive screening tools. Examples of these tools are the Montreal Cognitive Assessment (MoCA) or the Mini-Mental State Examination (MMSE). These names may be familiar to older adults and their caregivers since their administration is part of standard of care. Both tests have a maximum of 30 points, and cut-offs that indicate typical cognitive functioning, mild cognitive impairment, or dementia.. For example, Tan et al. (2015) proposed cut-offs for three age groups 65–79, 80–89, and 90 years or older. The MoCA cut-off scores for detecting Mild Cognitive Impairment are ≤25, ≤24, and ≤23, respectively, and for moderate cognitive impairments due to dementia they are ≤24, ≤21, and ≤18. Falling on one of these cut-offs may have implications on a person's autonomy since the result may be used for determining a lack of capacity for making decisions. When older adults and caregivers are on the journey of a possible cognitive decline faster than the typical aging process, they become familiar with the process of taking these screening tests and the potential consequences of the results. This may cause anxiety to the older adult being assessed, and anxiety affects performance in cognitive tests in older adults (Meltzer et al., 2017).

Serious mobile games (Figure 4.1) can be a low-cost, low-risk, enjoyable, and effective alternative to cognitive assessments. This is an emerging field receiving increasing attention from researchers and health practitioners. Serious games are developed using information and communication technologies and are designed for purposes other than entertainment (Wilkinson, 2016). These games have features that can make it easier and more enjoyable to monitor the cognitive status in older adults. First, mobile games do not look like an assessment tool, so anxiety associated with the traditional cognitive screening administration can be reduced. Older adults may enjoy the activity of playing mobile games regularly. This may provide a better idea of the actual cognitive status of an older adult. Second, some domains of cognitive functioning can be monitored on a regular basis using mobile games. Third, the older adults' performance can be automatically stored to track changes over time. Health care practitioners then can be informed of, and follow up on, such changes.

Researchers have investigated how standardized cognitive screening tool scores are related to the performance of older adults while playing mobile games. Findings include that older adults with higher performance in the games also have higher scores on the MMSE, so both outcomes are correlated (Tong et al., 2016; Wallace et al., 2017). In addition, algorithms used in mobile games correctly classified the older adults with typical cognitive functioning, mild cognitive impairment, and dementia (Valladares-Rodriguez et al., 2018), and predicted MMSE scores using the game scores (Jung et al., 2019). These results encourage the adoption of mobile games as a proxy of cognitive function and for detecting changes in cognitive functioning early, in a simpler and more enjoyable manner, making them more easily accepted by older adults than the traditional cognitive screening tools. Examples of platforms with these types of games are the ones created by Winterlight Labs (https://winterlightlabs.com) and by VibrantMinds (https://vibrant-minds.org/vibrantminds2/start) hosted by the University of Alberta, both developed in partnership with AGE-WELL.

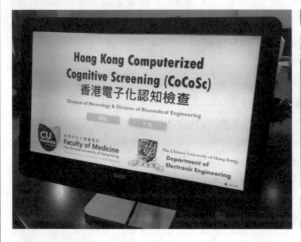

Figure 4.1: Serious games for cognitive assessments.

4.2.2   SELF-MANAGED HOME CARE MODELS

PERSONA

Edmund Bauer

- **Age:** 70
- **Gender:** Man, pronouns are he/him
- **Lives in:** Single family house with wife and youngest son, in Edmonton, Canada
- **Social circle:** Wife, 4 adult children, 2 grandchildren, and a few close friends

Signature interests

As a first generation Canadian, he remains engaged with the German community in Edmonton. He runs a family-owned construction company having no plans to retire any time soon. He is a proud Opa who spends most of his down time with his 2-year-old grandson and 4-year-old granddaughter.

"Despite me recently being diagnosed with COPD, I still need to work. My children are only getting started in their lives, and I would never want to be a burden to them. I want to do everything in my power so I can continue to work for as long as possible to help make ends meet."

Health
COPD, osteoarthritis.

Technology
Uses a smartphone for work, to keep in contact with family, and to surf the internet. Has Internet connection at home. Is willing to try other technologies as his COPD and osteoarthritis worsen.

Self-managed home care is an alternative model to traditional forms of home care where care recipients whose cognitive skills are either unaffected or affected to a minor extent, such as the case of Edmund in the persona above, are provided with more autonomy in determining what care they need and how that care should be delivered. In self-managed care, the government directly funds home care recipients, who can then purchase home care services from a provider of their choice. Numerous forms of self-managed care exist in various Canadian jurisdictions (Jamieson et al., 2021), including provinces such as Ontario, Alberta, and British Columbia, in addition to other countries such as England, France, Germany, and Australia.

Gotcare, a for-profit social enterprise, was established in 2018 to address the need for self-managed home care options, and it is the largest privately operated, self-managed home care provider in Ontario (Gotcare.ca, 2018). The company's mission is to make home care a "viable career for front-line workers," and to identify operational efficiencies in the sector (Gotcare, 2021). Their approach involves using an SMS-based technology to scan their network of home care workers and to match them with nearby care recipients. Gotcare sends the care recipient a short list of potential home care workers located nearby, and it is the responsibility of the care recipient or caregiver to choose a home care worker. Findings from Gotcare (https://gotcare.ca) were indicated by Jamieson et al. (2021) as a viable option within the home care sector, however further research is needed to ensure that the highest standard of care is delivered to care recipients, and to inform evidence-based policy decisions.

4.3 ETHICAL TENSIONS BETWEEN CAPACITY AND AUTONOMY

The autonomy of older adults is related to their ability to have the capacity to make decisions. In this part of the chapter, we discuss the ethical tension regarding the involvement of older adults in decision making.

4.3.1 INVOLVEMENT OF OLDER ADULTS AND PERSONS LIVING WITH DEMENTIA IN DECISION MAKING

The term *capacity* has been understood in different ways and at different times and places. It has been said that "there are as many different operational definitions of mental (in)capacity as there are jurisdictions." To further complicate matters, there are multiple terms, such as "mental capacity," "competence," and "decision-making ability" (and their opposites, including "mental incapacity" and "incompetence"), which are sometimes used as synonyms or near synonyms of "capacity," and used to make important conceptual distinctions. For instance, "competency" may denote "acceptable" behavior, the ability to take on tasks, standing in a legal proceeding, or a cognitive process of decision making (Law Commission of Ontario, 2020).

With the current view of autonomy being associated with individual autonomy (Harding, 2012), as highlighted earlier in this chapter, communication of how someone came to a decision is key. When an older adult or person living with dementia struggles with communication, there is a greater risk of the person's decisions to being ignored and, thus, their right to autonomy. This is irrespective of whether relevant tests have been conducted to determine whether the person has the capacity to make a particular decision. This focus on individual autonomy is problematic because they exist in social relations which can substantially influence their reasons for making particular decisions. The lives and environments of older adults and persons living with dementia are more likely to be limited by the relationships of care and dependency that support them (Bhatt et al., 2020).

There can be misunderstandings as relates to capacity, consent, and autonomy, especially among persons living with dementia (Neubauer and Liu, 2021). Diagnosis of dementia, for example, is often mistaken with capacity loss and does not take into consideration the progressive nature of the disease. This could be as a result of unstructured processes that lawyers follow when determining whether someone is eligible for guardianship, where many rely on gut feelings rather than making use of data from standardized assessments. This could be in part due to limited understanding of capacity and consent.

Pertaining to health care professionals, confusion also was found between lack of communication and the ability to consent; this can further contribute to abuse. For example, it is often assumed that if a person cannot communicate, the person is cognitively impaired and therefore others can make decisions for the person. As a result, some older adults who are fully capable to make decisions, such as those living with hearing loss or having suffered from a stroke, are at risk of being designated as being more impaired than they really are, which can result in them losing their ability to be their own decision makers.

It is also important to be clear that among persons living with a cognitive impairment, many do not lose capacity immediately upon their diagnosis, and that regardless of their level of capacity, there should always be a level of autonomy available—even if it is something as simple as deciding what they want to eat for breakfast or watch on TV (Rempala et al., 2020). For example, as noted by Jaworska (2017), the "potential for autonomy [is associated] primarily with the capacity to value, and well-being with living in accordance with one's values" (2017, p. 276). Being a "valuer" only requires the person to think that they are correct in wanting what they want and achieving what they want is tied up with their sense of self-worth. By these criteria, persons living with dementia are indeed "valuers," and can maintain partial autonomy through their ability to value. With this being said, while most persons living with dementia accept that they will eventually become dependent on their caregiver, regardless of the circumstance they still want their voices to be heard when involving any decision making that concerns them (Wolfe et al., 2020). Human rights frameworks, such as the FREIDA principles (Fairness, Respect, Equality, Identity, Dignity and Autonomy) (Butchard and Kinderman, 2019) and the Canadian Charter of Rights for People Living with Dementia (Figure 4.2), could serve as a guide to assist in the development of practical strategies which facilitate upholding people's rights on involvement in decision making on a day-to-day basis.

Canadian Charter of Rights for People Living with Dementia

As a person with dementia, I have the same human rights as every Canadian as outlined in the Canadian Charter of Rights and Freedoms. The following rights are especially important to me. I have the right:

1 To be free from discrimination of any kind

2 To benefit from all of Canada's civic and legal rights

3 To participate in developing and implementing policies that affect my life

4 To access support so that I can live as independently as possible and be as engaged as possible in my community. This helps me:
· *Meet my physical, cognitive, social, and spiritual needs*
· *Get involved in community and civic opportunities*
· *Access opportunities for lifelong learning*

5 To get information and support I need to participate as fully as possible in decisions that affect me, including care decisions from the point of diagnosis to palliative and end-of-life care

6 To expect that professionals involved in my care are:
· *Trained in both dementia and human rights*
· *Held accountable for protecting my human rights including my right to get the support and information I need to make decisions that are right for me*
· *Treating me with respect and dignity, offering me equal access to appropriate treatment options as I develop health conditions other than my dementia*

7 To access effective complaint and appeal procedures when my rights are not protected or respected

Based on: Alzheimer Society of Canada

Figure 4.2: Adaptation of the Canadian Charter of Rights for People with Dementia. Courtesy of Hector Perez, Postdoctoral fellow at the University of Waterloo.

Such tensions in capacity among older adults and persons living with a cognitive impairment has led to a concept known as person-centered autonomy (Harding, 2012; Wolfe et al., 2020). Rather than requiring that autonomy be sacrificed in the name of best interests, a person-centered approach to autonomy allows welfare to be considered as essential to a person's autonomy and draws attention to the complexities of the power relations that a person living with dementia might be experiencing. Respecting autonomy through a person-centered approach does not require that any decision made by a person with limited capacity is carried out, but rather it requires significant input from others. Therefore, the person's wishes and feelings are not excluded or overridden, even if the outcome is the same. Enabling the person with some elements of decision making is the focal point of this form of autonomy (Harding, 2012).

FIND OUT MORE

1. Canadian Charter of Rights for People Living with Dementia: https://alzheimer.ca/en/take-action/change-minds/canadian-charter-rights-people-dementia

CHAPTER 5

Advance Care Planning

WHAT IS IN THIS CHAPTER?

This chapter introduces, describes, and discusses the role of advance care planning on autonomy among older adults and persons living with cognitive decline. We will then identify the potential role of solutions such as self-sovereign identity on autonomy among this population. Finally, we will discuss the ethical tensions as they relate to advance care planning and autonomy.

5.1   ROLE OF ADVANCE CARE PLANNING ON AUTONOMY

In preparation for the eventual, significant loss of autonomy among older adults and persons living with cognitive decline, **advance directives or advanced care planning** have been seen as "an effective means of extending an individual's autonomy from their current self, as an individual who has decisional capacity, onto their future self, who lacks it" (Walsh, 2020, p. 54). It does this by inviting the person to choose who they would want to make important decisions (e.g., health care decisions) for them when they are no longer able to make such decisions for themselves. This would also include instructions about what decisions they would want the designated person to make for them and/or to describe their values and beliefs to guide the decision making about what they would have wanted in each situation. It is recommended that an advance directive is created when the person is well due to the potential for sudden changes in their health that would impact their capacity, such as getting into a serious car accident (Health Law Institute, n.d.). In fact, the Alzheimer Association (n.d.) also expressed the importance of such documents. From the context of the LGBTQIA2 population, such as that described by Marni, whose persona was featured in the first part of this book, advance care planning is vital as it attempts to redress the exclusionary actions of governments (de Vries, 2009), specifically in the context of the complex networks LGBTQIA2 persons live in, where, for many, some (or no) family members are included in their care (Muraco, 2006).

Despite its stressed importance, as noted by Hamel (2017) and Newcomb (2020), only about half of Americans have had end-of-life conversations with loved ones, and only 27% with formal documentation. In fact, in a study by Snyder et al. (2013), only 43% of patients were provided with advance care planning by their clinicians (Snyder et al., 2013). Clearly, there have been several types of barriers to these types of discussions with the public, and they include difficulties with legal components of the advance care planning process, lack of knowledge, and concerns over the timing of advance care planning conversations (Beringer et al., 2021; Boddy et al., 2013).

In addition to advance directives, there are other options to assist someone living with dementia or an older adult in preparation for, or when they no longer have the capacity to make important decisions for themselves. This can include a **power of attorney** which grants legal authority to an individual to make decisions on behalf of another. In Canada, a power of attorney is a legal document that allows a person's designated attorney to manage their finances and property on their behalf when they lack the capacity to manage their own affairs (Government of Canada, 2016). **Trusteeship** is needed only when the adult has limited financial assets and only needs help managing income from a government program or pension, such as Old Age Security. An adult with a power of attorney agreement does not need a trustee, however some agreements end when an adult loses capacity to make decisions (Government of Alberta, 2021b).

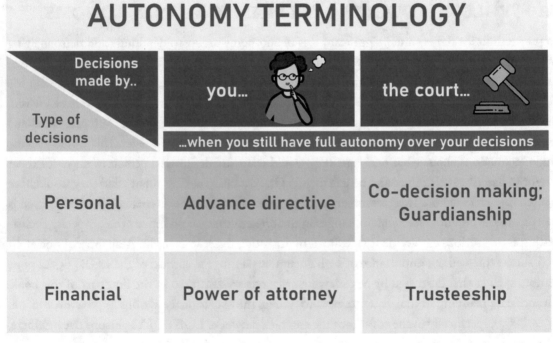

Figure 5.1: Summary of autonomy terminology. Courtesy of Hector Perez (Postdoctoral fellow, University of Waterloo).

Guardianship is an option available when the person does not have a power of attorney or advance directive already in place. Guardians are appointed based on professional options and a decision of the court or any other legal authority once a person is declared incompetent to handle their own affairs (Doron, 2003). The court then transfers the responsibility for social activities, employment, legal proceedings, living arrangements, and medical decisions to *the guardian*. Which decisions *the guardian* is responsible for is dependent on the court's decision (Government of Alberta, 2021c). Guardianship can take many forms. The traditional form gives *the guardian* full authority to make decisions for the adult who has lost legal capacity, and partial or limited guardianship, where the court lists specific powers over which *the guardian* has authority. Partial or limited guardianship has served as an attempt within Canadian law to respect the adult's autonomy, encourage independent decision making, and potentially aid the adult in regaining functional capacity (Holmes, 1996). **Co-decision making** (also known as supported or assisted decision making) is another model of guardianship within Canada where a co-decision maker assists the adult in making decisions, rather than making decisions for them. While this model has become more attractive (Dementia Alliance International, 2016), co-decision-making legislation is currently available only in Saskatchewan, Manitoba, and the Yukon (Burningham, 2009).

5.2   SOLUTIONS TO SUPPORT ADVANCE CARE PLANNING

As highlighted in this chapter, the concept of advance care planning can be used as a means of ensuring one's wishes are fulfilled in the event that they lose their autonomy. There has been an increase in the number of strategies that assist in this process. This can include high tech strategies such as self-sovereign identity and blockchain. While this part of the chapter does not encompass all tools and strategies that are available in the area of advance care planning, it provides the reader with examples of emerging solutions.

5.2.1   SELF-SOVEREIGN IDENTITY

PERSONA

Jamie Stirling

- **Age:** 56
- **Gender:** Man, pronouns are he/him
- **Lives in:** Single-family home with his wife, in Listowel, Ontario
- **Social circle:** Active volunteer in his community

Signature interests
Physical activity, golf, community service, entrepreneurship.

" As a retired police officer serving the community, when it comes to using my experience and knowledge to the business world, it is a whole different culture—it is a collision that makes it difficult to find investors in my project."

Health
Good health conditions.

Technology
Integration of technology and data for the safety of the citizens.

Self-sovereign identity (SSI) (Figure 5.2), an identity management system which can allow an individual to fully own and manage their digital identity (Mühle et al., 2018), has the potential to provide older adults, such as Edmund and Gladys who were introduced earlier in this book, and persons living with dementia, such as Roger and Michaela, with more control over who owns and controls their digital data. Under the SSI model, people can prove their identity and get access to records, such as their health data, without needing to go through an intermediary such as government-issue identification or passwords. Blockchain is one way to implement a system with SSI (Mühle et al., 2018) by providing a person-to-person network where a transaction occurs between authenticated users. SSI may be a solution that can benefit persons living with cognitive decline such as dementia by providing relevant, secure information that first responders, such as Jamie in the persona seen in this section, and care providers can use if something happens to them. The same could be said for older adults who do not have a cognitive impairment, where they can gain easier and quicker access to their health information. Older adults and persons living with cognitive decline have the potential to also share their anonymous data with researchers, policy makers, and program planners to use in determining resource allocation and program planning and can control what information their designated guardian has access to. The concept of SSI, however, is in its infancy in Canada, with banking systems being the first to implement this technology within British Columbia, Alberta, and Ontario (Digital Canada, 2019). Because of the novelty of this concept, few studies evaluating SSI exist, and the potential use for SSI has yet to be investigated within a health care context.

When an older adult or someone living with dementia no longer has the capacity to consent and make decisions for themselves, who becomes the primary owner of their data? As noted by the Sovrin Foundation, guardianship is an essential component of the Sovrin Governance Framework. Like the traditional forms of guardianship highlighted earlier in this chapter, it is designed to account for these instances when one cannot be fully self-sovereign (i.e., have limited decision-making capacity), such as the case of those living with a cognitive impairment. While the preliminary structure of guardianship within the context of SSI has been identified, as seen in the Sovrin whitepaper about guardianship (Sovrin Guardianship Task Force, 2019), it has yet to receive feedback from the general public in terms of its feasibility.

SELF–SOVEREIGN IDENTITY
WHAT IS IT?

 Security: Online identity systems are insecure. *Blockchain* adds a layer of security and allows you to share only the necessary data for a transaction or interaction. This system is a single way to verify your identity.

 Privacy: Privacy is a critical feature of self-sovereign identity because it gives you control. Your digital data is yours and you can choose who to share it with.

 Ownership & Control: "Self-sovereign" means sole ownership of your digital identity, you control how your data is shared and used.

 **Digital Identity**: Your data forms your identity. Self-sovereign identity uses *blockchain* technology to securely register your digital identity. Your data is stored in your electronic device and used to verify your identity like a driver's license.

 **Removal of Centralized Authority**: Your personal information is owned and controlled by you. Therefore, you only need an electronic device to access your personal data instead of going through a third party like a government website.

Figure 5.2: Summary of Self Sovereign Identity. See video for overview (https://www.youtube.com/watch?v=0WicIm8x_GY&feature=youtu.be). Courtesy of Hector Perez.

5.2.2   ADAPTABLE DECISION-MAKING SUPPORT

Technologies should accommodate changing decision-making capacity (Burmeister, 2016). This has been variously described as ambient intelligence, embedded systems, and, more recently, deliberative assistive technology devices (Teipel et al., 2016), such as Gerijoy (http://www.gerijoy.com/senior-living.html). An example is a company that uses ambient intelligence avatars and human relational telehealth products to work with older adults and persons living with dementia. Their system depends on input from remote caregivers, making it adaptable to the changing circumstances of the person to whom care is being provided. In futuristic scenarios, such as those proposed by Rogers and Mitzner (2017), the passive monitoring of a person's everyday activities, physiological status, and emotional state via sensors and voice detection could also be used to predict functional changes which in turn could be used to adjust the technology requirements and determine if additional services are required.

5.2.3   TOOLS TO ASSIST ADVANCE CARE PLANNING

Online Resources

Since the onset of the pandemic, in the United States there has been a tenfold increase in the number of people requesting advance directives (Aleccia, 2020). Despite this, limited resources exist. Five Wishes (https://fivewishes.org), Respecting Choices (https://respectingchoices.org), and The Conversation Project (https://theconversationproject.org) are all tools that could be used to assist in this process. In Canada, Dying with Dignity Canada's Advance Care Planning Kit (https://www.dyingwithdignity.ca/download_your_advance_care_planning_kit), resources provided by Dalhousie University's Health Law Institute (http://eol.law.dal.ca/?page_id=231), and the My Wishes, My Care Planning Tool (https://www.advancecareplanning.ca/news/my-wishes-my-care-starting-the-conversation-january-20-27-2021/) could also be helpful. It is important to note, however, that advance care planning online may not be recommended for everyone, due to the risks of not seeking adequate legal representation.

Online Advance Directive Registries

In the United States, there are a number of online "registries" where people can load a drafted advance directive into an online registry database and get a card that indicates how people can access that advance directive in an emergency. For example, there are government-run advance directive registries in Arizona, California, Idaho, Maryland, Michigan, Montana, and Washington. There are also several private and state advance directive registries (https://www.americanbar.org/groups/law_aging/publications/bifocal/vol_37/issue_6_august2016/tour-of-state-advance-directive-registries/).

Within the Canadian context, unfortunately there is not a consistent and transferable mechanism for all care providers to share information about advance care planning and ensure that conversations regarding advance care planning continue across all care settings through an individual's care journey (Hawryluck, n.d.; National Seniors Strategy, n.d.). Small steps, however, have been made in recent years, with provinces such as Quebec creating a provincial registry (https://www.ramq.gouv.qc.ca/en/citizens/health-insurance/issue-directives-case-incapacity) for documenting preferences about refusing or accepting specific medical interventions if the person ever becomes incapable of giving consent (Gouvernement du Quebec, 2020). This registry, however, reduces advance care decisions to a limited set of choices and does not adequately reflect the full range of values people might have about end-of-life preferences (Bernier and Regis, 2019). To facilitate advance care planning, it is recommended by the National Seniors Strategy (n.d.) that proper documentation protocols are needed to facilitate accessibility of advance care planning in many care settings.

Advance Care Planning Applications

As highlighted by the website Tech-enhanced life (https://www.techenhancedlife.com), there has been an emergence of online applications for managing advance directives. In a review conducted by McDarby et al. (2020), nine mobile applications that facilitated each stage of the advance care planning process were included. Of these included applications, six permitted users to document a preferred decision maker, and six offered a mechanism to distribute and share advance care planning documentation. Applications included MedStar CR, My Dot Mediq (http://www.mydotmediq.com), BIDMC Health Care Proxy (https://www.bidmc.org/patient-and-visitor-information/patient-information/preparing-for-your-stay/ma-health-care-proxy/bidmc-health-care-proxy-mobile-app), and My Directives (https://mydirectives.com).

It is important to note, however, that the included applications have incomplete advance care planning content and lack essential design features that promote user friendliness. In addition, several of these mobile applications are not truly mobile applications, but instead are platforms with links to pre-existing advance care planning websites. This raises usability concerns as applications that link to websites may not load when the user is not connected to the Internet. As a result, there remains a strong need to develop rigorous, well-designed applications grounded in the advance care planning research evidence base. Key recommendations from this review include: (1) the need for basic information on advance care planning embedded within the application, including that it is a process of behavior change that can be broken down into a series of steps; and (2) a description of what the advance care planning process the application is intended to support (e.g., supporting decision making, providing information, etc.).

5.3   ETHICAL TENSIONS BETWEEN ADVANCE DIRECTIVES AND AUTONOMY

In this part of the chapter, we discuss the main ethical tension on advance directives.

5.3.1   ADVANCE DIRECTIVES

There are varying degrees of legal strength on these documents (Vezzoni, 2005). In the United States, Belgium, and the Netherlands, for example, advance directives are legally binding on physicians (De Boer et al., 2010). In countries such as Germany and Norway, these documents have a "weak" legal status. In these countries, while advance directives are viewed as having some legal status, they do not hold decisive moral weight in medical decision making (De Boer et al., 2010). In some countries, such as Japan, advance directives are given no legal status (De Boer et al., 2010).

While advance directives could be used as a means to support those that have lost their ability to remain autonomous, they leave out the person's wishes after the directive has been created. For example, as highlighted in a case study by Sokolowski (2018):

> At the age of eighty-five, Mrs. Black received a diagnosis of mid-stage dementia. Due to the progression of her dementia, Mrs. Black often struggled to recall the names and faces of family members. Nevertheless, she was noted by her nurses at the residential aged care facility she lived at as being an exceptionally happy woman. She took joy in her daily activities, particularly watching birds pass by through the window. While in care, Mrs. Black developed a serious bacterial infection. She had an advance directive stipulating that if she were ever to suffer an illness which resulted in her inability to recognize her family members, she would not wish to receive any medical treatment to prolong her life. Consequently, her son insisted that her advance directive be followed and so the directive was implemented. She died shortly after. The entire medical team report feeling "devastated." Mrs. Black's nurse Lili surmised, "when Mrs. Black wrote the advance directive, she probably did not consider the idea that she could have dementia and still enjoy a good quality of life" (Sokolowski, 2018, pp. 45–83).

Like the excerpt above, there is contradictory evidence to support that a person's change in preferences throughout the experience of an illness or disease is insufficient reason to assume post-onset preferences are inauthentic, non-autonomous, or bear less moral weight (Walsh, 2020). Many persons living with dementia, for example, undergo a cognitive transformative experience (Paul, 2016). Given that the transformative experience changes the way the person thinks about existing beliefs, values, and preferences, the implications of dementia cannot be fully considered by someone who is in the process of creating an advance directive (Walsh, 2020). Therefore, within the context of persons living with dementia, an advance directive may not reflect individuals' current values and interests. Taking this into consideration, the most appropriate way may be to advocate for a medical decision-making process that supplements a person's remaining autonomy via community engagement. This could include: (1) establishing common attitudes regarding the individual's current preferences and critical interests; (2) recognizing that the person has certain rights; and (3) contributing actions to the person's life which are in accordance with their values (Rempala et al., 2020). By doing so, caregivers can respect the autonomy of a person living with dementia and older adult by supplementing and reconstituting the person's diminished autonomy and fulfill critical interests without sacrificing current values and quality of their current life experience (Rempala et al., 2020).

The inclusivity of advance care planning also remains in question especially among LGBTQIA2 populations. In a study by Beringer et al. (2021), resources such as the My Wishes, My Care Planning Tool were found to be heteronormative in nature, focused primarily on the involvement of the biological family, paid less attention to inclusive language such as "partner," and did not display "more definitive photos of LGBTQIA2 couples." More inclusive versions have therefore been expressed as being warranted that take into consideration these limitations (Beringer et al., 2021).

Box 5.1: Health Care Consent, Aging and Dementia: Mapping Law and Practice in British Columbia

The Canadian Centre for Elder Law in partnership with the Alzheimer Society of British Columbia led a project which examined the law, policy, and practice of consent to health care in the context of aging and dementia. It involved extensive comparative legal research on informed consent and interrelated areas of the law, as well as community and key stakeholder consultation (Canadian Centre for Elder Law, 2019). The work was informed by an expert interdisciplinary advisory committee and culminated in a report identifying areas for law and practice reform and a set of plain language resources on health care consent rights for people living with dementia and their families.

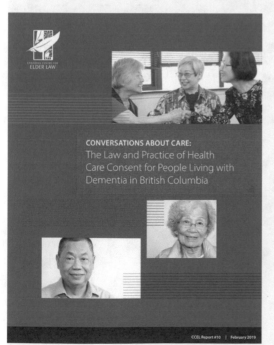

CONVERSATIONS ABOUT CARE:
The Law and Practice of Health Care Consent for People Living with Dementia in British Columbia

CCEL Report #10 | February 2019

"What happens is that there's a belief that the consent happens when you cross that threshold into care...Without realizing that, every step in that care, everything you do that's different, needs a consent"

- Health authority staff

Concerns identified by health care professionals and community:

1. There is not enough time to have discussions on health care decision making.
2. Some health care providers do not understand the law.
3. Everyone requires better support with decision-making.
4. Some people face unique barriers to health care decision making.
5. People do not know what to do when they disagree with the actions of health care providers or family.

Recommendations

1. Improve health care decision-making laws in BC.
2. Support the best practice of health care providers.
3. Address barriers to informed consent.
4. Enhance access to legal information and assistance.

FIND OUT MORE

1. Health Care Consent, Aging and Dementia: Mapping Law and Practice in British Columbia: https://www.bcli.org/wordpress/wp-content/uploads/2019/02/HCC_report-Final_web_Mar-29-2019.pdf

CHAPTER 6

Risk

WHAT IS IN THIS CHAPTER?

This chapter describes the role of risk on autonomy among older adults and persons living with dementia. Following this, solutions such as guidelines and educational programs will be described and how they influence risk and autonomy. Finally, the ethical tensions between risk and autonomy will be described and how they could potentially affect the autonomy and overall quality of life of this population.

6.1 RISK: THE ABC

Risk is a combination of two components, the probability of occurrence of harm and the consequences of that harm; in other words, how severe the harm might be (International Organization for Standardization, 2007, p. 4). **Harm** is any physical injury or damage to the health of people, or damage to property or the environment (International Organization for Standardization, 2007, p. 4). A hazard is a potential source of harm. Normally when we interact with technologies and or processes, we are exposed to hazardous situations. **Hazardous** situations are those "circumstances in which people, property, or the environment are exposed to one or more hazard(s)" (International Organization for Standardization, 2007, p. 4). So, there is a relationship between "hazard" and "hazardous situation" that lead to risks. A hazard cannot result in harm until such time as a **sequence of events** leads to a hazardous situation. Table 6.1 provides examples on how a hazard can be transformed into a hazardous situation and produce harm by a sequence of circumstances. Figure 6.1 shows a representation of the relationship of hazard, sequence of events, hazardous situation, and harm.

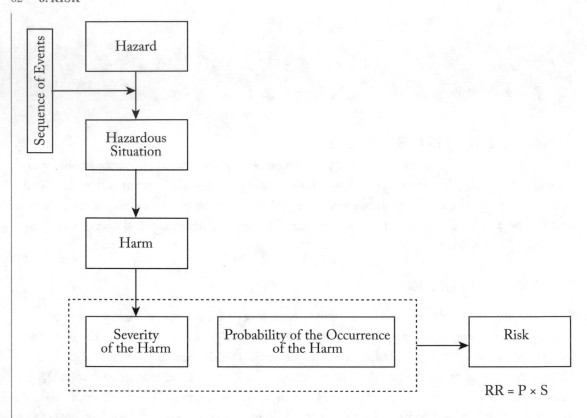

Figure 6.1: Relationship of hazard, sequence of events, hazardous situation, and harm. (Modified from International Organization for Standardization, 2007, p. 4.)

Table 6.1: Relationship between hazards, foreseeable sequences of events, hazardous situations, and the harm

Hazard	Foreseeable sequence of events	Hazardous situation	Harm
Thermal energy (low temperature)	Older adult was unintentionally unattended by a nurse	Older adult went out of the facility and, as a result, he/she is exposed to low temperatures (environmental factors)	Death

6.2   ROLE OF RISK ON AUTONOMY

The management of risk has become increasingly central to the practice, assessment, and decision making among older adults and persons living with dementia (Clarke et al., 2011; De Courval et al., 2006; Manthorpe, 2003). Historically, the notion of risk was recognized as being neutral, referring to the probability of an event occurring. Therefore, risk could be either something good or bad which could involve loss or gain, such as the risk of winning the lottery (Lupton, 1993, 1999). However, in modern society the meaning of risk has been transformed from being a neutral term into something that is entirely negative and dangerous (Douglas, 1990). High risk equates to a lot of danger in modern western cultures (Douglas, 1990); furthermore, inherent in this notion of risk is the probability of harm or undesirable experience. It is challenged that this **perceived risk** is created by society and mediated by social and cultural processes (Lupton, 2005). It has moved to the point where we now live in a world of fear, where we concern ourselves with anything and everything from the air that we breathe to the food that we consume (Lupton, 1993).

Such perceptions about risk as harm are embedded within health care practice where, in the case of older adults and persons living with dementia, risks are related more to their perceived vulnerability than to potential harm to others (Manthorpe, 2003). In the case of Edmund, as described in Chapter 4, despite him being aware that his COPD is a result of his 45+ years of smoking, he continues to live with risk by smoking ½ pack/day. There are some exceptions to harm only affecting the individual, such as leaving the stove on or worsened driving skills (Man-Son-Hing et al., 2007). The main risks associated with this population, however, relate to the greater likelihood of physical harm to themselves (Manthorpe, 2003). This perceived vulnerability has led to the view that as one gets older or as someone progresses along the dementia continuum, they require constant supervision/surveillance, and must be restrained through restriction of activities, and it denies an older adult or person living with dementia the right of choice, self-determination, self-esteem, and respect (Macdonald and Dening, 2002; Titterton, 2005). This has inevitably resulted in poor standards of care, and at its worst can lead to the abuse of an older adult or person living with dementia (Everitt et al., 1991; Macdonald and Dening, 2002; Titterton, 2005).

In most cases, this perceived risk among older adults and persons living with dementia can be exaggerated, with emotions playing a role. People tend to judge risk by how they think and feel about it, placing a reliance on such feelings to guide their judgment and decision making (Slovic and Peters, 2006). For example, following a tragic fall of an older adult while out and about in the community, what was once perceived as a low-risk activity, might quickly be seen as high risk among family caregivers due to heightened feelings of fear following the incident. As a result, it can mislead individuals responsible for the care of an older adult or person living with dementia in developing drastic restrictions and precautions in situations that may not be necessary (Slovic and Peters, 2006). Risks that have low probability but high consequence, for example, may be feared out of proportion (Manthorpe, 2003), such as that of wandering in persons living with dementia.

In addition to risk perception, **harm reduction** is another important concept as it pertains to risk and autonomy among older adults and persons living with dementia. Harm reduction can include any strategy that is designed to reduce harm or the probability of something negative happening, without requiring the complete cessation of the risky behavior. Examples of harm reduction in other disciplines include providing free condoms at university campuses or safe injection sites for drug users. Doing so would enable balance between the caregiver and care recipient, who may have different perspectives on the benefits and risks of specific behaviors and activities (Egan et al., 2017). It would also assist in changing how risk is perceived; from something seen as dangerous to one that is a neutral term. This would enable transformation of the current practice of safety maintenance to one with a more positive "**ethics of care**," which could be characterized as one that emphasizes the importance of relationships between the caregiver and care recipient, provides a commitment to effective communication, and one that appreciates the uniqueness of individuals in unique situations (Hughes and Baldwin, 2006).

6.3   ROLE OF SOLUTIONS ON RISK AND AUTONOMY

As highlighted in the previous part of this chapter, the concept of risk in relation to autonomy among older adults and persons living with dementia is complex and, like capacity, can have a substantial influence on this populations' ability to make decisions. There has been an increase in the number of available strategies in recent years, from guidelines to technology repositories, and various education and awareness tools such as those offered by the Alzheimer Society of Ontario's Finding Your Way® Program that can be used to help older adults and persons living with dementia. While this part of the chapter does not encompass all tools and strategies that are available in the area of risk and autonomy, it will provide the reader with a taste in terms of the possibilities that are out there.

6.3.1   GUIDELINES AND REPOSITORIES TO SUPPORT DECISION MAKING OF RISK MANAGEMENT STRATEGIES

To assist in reducing common risks among older adults, **risk management strategies** are commonly used. Risk management strategies can be defined as a "systematic application of management policies, procedures and practices to the tasks of analyzing, evaluation, controlling and monitoring risk" (International Organization for Standardization, 2007, p. 4). In recent years there has been a significant increase in the number of available risk management strategies for older adults and persons living with dementia. For example, as highlighted through Michaela's persona in Chapter 2, and will be described in Chapter 12 of this book, GPS could be used to reduce the risks associated with getting lost among persons living with dementia. By wearing a locating device, a caregiver can pinpoint the approximate location of their loved one who is lost, which would help searchers find and return the lost person home faster (Freiesleben et al., 2021). Among older adults who live with a high fall risk, fall detection devices could be integrated whereby caregivers are notified when they fall. This could subsequently reduce the time it takes for the older adult to get help if they cannot get up and are alone (Güttler et al., 2018). While they may provide promise for enhanced quality of life for this population, information describing these available strategies remains inconsistent and diverse, contributing to the challenges that caregivers, persons living with dementia, and older adults face when understanding how these strategies work, and how to choose a strategy that best suits their needs. Over the years various guidelines and repositories have been developed to help provide more trustworthy information on strategies including technologies. The Canadian Guideline for Safe Wandering, for example (Box 6.1), was created in collaboration with a wide range of stakeholders, including persons living with dementia and caregivers. The aim of this guideline was to simplify and summarize available strategies to manage the risks associated with critical wandering. While this guideline in its paper and web-based formats has been successful since its initial launch in 2018, a gamified digital version would enhance its adoption. As noted by Brown et al. (2013), for example, traditional guideline development and dissemination occurs through written publications, which face the risk of being outdated before they are published and available to mainstream populations. In addition, traditional guidelines may not be as engaging as alternative versions. Interactive versions of such guidelines that use gamification components, such as "Let's Talk about Dementia" by the World Health Organization (https://www.paho.org/lets-talk-about-dementia/warning-signs/), would improve the end user's ability to remain engaged and motivated to use the developed tool (Deterding et al., 2011).

Other useful resources are health tech libraries, which serve as online platforms that contain a variety of health-related technologies that can be used to manage common safety risks among older adults and persons living with dementia. For example, these can include websites such as Grips and Gadgets (https://gripsandgadgets.com), CanAssist (https://www.canassist.ca), the Alzheimer's Store (https://www.alzstore.ca), in addition to the Locating Technology Repository (https://tech.findingyourwayontario.ca) that is a part of the Finding Your Way® Program with the Alzheimer Society of Ontario.

Box 6.1: Canadian Guideline for Safe Wandering

Person Living with Dementia Version

RISK OF GETTING LOST

Low Risk	Medium Risk	High Risk	Unplanned Absence
☐ Initial signs present or you have a diagnosis of dementia ☐ You live with family and have them around 24/7 ☐ You go outside without having someone go with you	☐ You live with family but are normally home alone or go out with friends/family away from the home ☐ You exercise when you are stressed ☐ You get overwhelmed or anxious frequently	☐ You live at home alone ☐ You leave your home or go for walks alone ☐ You walk away from your friends/family when you are out in the community with them	☐ You are lost within the community

STRATEGIES

Low Risk	Medium Risk	High Risk	Unplanned Absence
• Talk to your local Alzheimer Society or care provider • Develop a plan of preventative strategies • Use identification strategies (e.g., ID tags or bracelets)* • Register yourself under a vulnerable persons registry if available in your area*	• Exercise with a partner or use a locating technology • Create list of where you used to live/work and keep list at home • Have someone look out for you more (i.e., locating devices or have a friend check in with you daily if you live alone)	• Look into you and your care partner using a locating technology • Start a buddy system • Seek community supports (i.e. home care)	• **Call 911 or go to a local business to ask for help** • Use google or apple maps to redirect yourself and keep your home address saved on your phone or on a card that can be kept in your wallet or jacket • Have a friend or family member you can call

Need to apply education and proactive strategies as soon as possible so can still encourage **safe** or **purposeful** wandering

Can transition to lesser or greater levels of risk at any moment

NOTE: Strategies with a star (*) next to it could be applied to any risk level

Introduction

The Canadian Guideline for Safe Wandering is a comprehensive, easy-to-use guideline that offers proactive strategies to reduce the risk that someone with dementia will get lost. It is available in three versions: one for care facilities, one for community settings, and one for persons living with dementia. The guideline was developed in collaboration with persons living with dementia, caregivers, health professionals, first responders, community organizations, and Alzheimer Societies. The guideline serves two purposes: (1) to assist persons living with dementia and caregivers in choosing proactive strategies that work best for them; and (2) to serve as a discussion piece to be used among health professionals and community organizations.

Explanation

The two-page guideline is broken down by risk level of getting lost (i.e., low, medium, high, and unplanned absence). Within each risk level, a series of check boxes are provided to assist users in identifying what level of risk the person with dementia is at for getting lost. Below this are suggestions of proactive strategies to use, with further information on these strategies being provided on the back page. Messaging on the bottom of the guideline such as "can transition to lesser or greater levels of risk at any moment" was added to highlight that lost incidences are unpredictable and can occur even when the person with dementia is classified as being at a low risk of getting lost. Because of this, persons living with dementia and caregivers are encouraged to have multiple strategies in place. The message "need to apply education and proactive strategies as soon as possible so can still promote safe or purposeful wandering" was also added. As highlighted from Neubauer's work, this form of wandering, such as going out for walks and engaging with their community is important for persons living with dementia. Therefore, the guideline promotes this behavior, provided that proactive strategies are in place to allow them to do so safely.

6.3.2   EDUCATION

Timing of risk management education and awareness

While strategy guidelines and repositories as described above can be powerful in reducing safety risks among older adults and persons living with dementia, most strategies are implemented retrospectively (i.e., after an incident that threatens the person's safety, such as getting lost, has already happened). As a result, most older adults, persons living with dementia, and/or caregivers are driven to implement strategies quickly when they are at a heightened emotional state caused by the incident. Heightened emotional states in turn can lead to a significant loss of autonomy such as institutionalization, for the sake of keeping the person safe. As highlighted in the previous part of this chapter, to avoid this sudden shift in risk perception, a balance between risk and safety is emphasized by Robinson et al. (2007) and Neubauer and Liu (2021) (Box 6.2) and can be done by implementing strategies *proactively* (i.e., *before* an incident that threatens the person's safety has happened). Balance of risk and safety, as a concept, needs to be emphasized further to ensure that quality of life of this population remains intact. While wandering among persons living with dementia, for example, is a behavior that is of concern due to the risk of getting lost and its associated outcomes, not all forms of wandering lead to a missing incident (Dewing, 2006). Wandering should not be discouraged entirely due to the benefits associated with this behavior, such as providing the person with dementia with a sense of identity and engagement with his or her community (Brittain et al., 2010; Dewing, 2006), coping with stressful conditions, and as an ability to exercise (Bonifazi, 2000).

Box 6.2: Goldilocks Principle on Dementia and Wayfinding

The "**Goldilocks Principle on Dementia and Wayfinding**" highlights the need for the perception of risk of critical dementia-related wandering being "just right" among caregivers, persons living with dementia, and health professionals. Within this principle, the different levels of risk perception are proposed, shedding light on the limitations that come with each end of the spectrum (i.e., no risk perception ensures independence while sacrificing the safety of the person with dementia, while high-risk perception, ensures safety while sacrificing their independence). This principle proposes that future strategies and education directed to involved stakeholders must strive toward addressing an "optimal risk perception" where all parties involved see the risk of persons living with dementia getting lost, resulting in adoption of proactive strategies that enable some level of independence for the person with dementia while still ensuring that they remain safe within their home of choice (i.e., early adoption of monitoring devices). This principle was developed with the intention of including it in a conceptual framework that would encompass proactive strategies to mitigate the risks associated with getting lost. It is anticipated that an understanding of the concept of "optimal risk" captured within the Goldilocks Principle, would result in adoption and use of risk mitigation strategy guidelines.

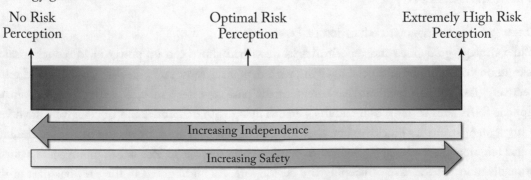

Education Programs

As highlighted above, education and awareness programs can assist in the incorporation of proactive strategies to reduce safety risks among persons living with dementia and older adults. The Finding Your Way® Program's Living Safely with Dementia Resource Guide and main website (http://findingyourwayontario.ca) is an excellent example of this. In partnership with people living with dementia, caregivers, and other key stakeholders, the purpose of the Finding Your Way® Program is to provide tips and strategies on various safety-related topics to help people living with dementia live safely in their day-to-day lives. It also provides strategies for safety planning, home safety (e.g., lighting, planning for emergencies, use of dangerous tools/appliances), sleep habits, staying social, staying active, driving, travel, medication, nutrition and food, living alone, and assessing the living environment. Other guidelines can include toolkits developed by the World Health Organization such as "Let's Talk about Dementia" and resources available through the International Consortium on Dementia and Wayfinding (International Consortium on Dementia and Wayfinding, n.d.).

It is important to note, however, that too often guidelines, repositories, and education programs that focus on helping persons living with dementia and older adults mitigate risk are developed for family caregivers (Neubauer, 2019). Thus, there are few resources available for persons living with dementia and older adults to be active agents in their own care. It was suggested by many that if the work is not catered specifically to the population of interest, necessary education and awareness strategies are often ignored. For example, one person living with dementia that participated in the study conducted by Neubauer (2019), stated, *"When it's not addressed to me, I'm not going to pick it up. There needs to be something that attracts my attention"* (2019, p. 225). This finding is similar to other literature that emphasizes sustainability and uptake of educational tools and resources (Beer et al., 2011; Nayton et al., 2014). With the rising number of older adults and persons living with dementia, and a greater number living alone (Eichler et al., 2016), more approaches of this nature need to be made to ensure autonomy is maintained, and the implementation of technologies, strategies, and education programs to manage risky behaviors and activities are adopted.

6.4   ETHICAL TENSIONS BETWEEN RISK AND AUTONOMY IN HEALTH CARE

Box 6.3: Tensions between risk, autonomy, and health care settings. Author: Adebusola Adekoya (Ph.D. student, University of Waterloo).

Health professionals have the ethical duty to respect the autonomy of patients and take actions to minimize risk and maximize their safety. However, the patient's need for autonomy may conflict with the goals of care or care guidelines (Haddad and Geiger, 2020). Supporting patient autonomy can be challenging at times, especially when the patient's choice involves risk. Further, risk is not only difficult to define but also difficult to evaluate and communicate

(Hunt and Ells, 2011). Risk is associated with harm, especially for people living with dementia. Harm, which can be an outcome of taking or not taking a risk, can have consequences, including legal repercussions (Ibrahim et al., 2019).

An individual's perception of risk is influenced by factors such as past experiences, fears, personal values, and worldview and may change over the course of one's life. Because of differences in how risks are perceived, behaviors and actions interpreted as "risky" by health professionals may be perceived as constructive and personally relevant to older adults (Rush et al., 2012). For example, "wandering" can be a form of physical exercise and relieve residents living in long-term care homes from boredom (Adekoya and Guse, 2019). However, health professionals may feel obligated to prevent residents from wandering or control the behavior to mitigate risks such as falls and going missing. Also, health care facilities that fail to take precautions to keep residents safe may be found liable for any wandering related injuries or death (Meek, 2014; Mulligan and Levin, 2000). Therefore, resident safety may be viewed as more important than autonomy due to perceived risk of litigation.

To address the tensions between risk and autonomy in health care, it is imperative that health professionals understand how patients perceive risk. There must be open and honest communication between health professionals and patients and their caregivers regarding risk and its consequences. Risk agreements or negotiated risk agreements that outline patients' decisions on the acceptable level of risks and consequences they are willing to take and live with and how health professionals can support their decisions should be made (Ibrahim et al., 2019). Being able to make an informed choice about whether to take risk or not should be the fundamental right of all patients, including older adults. This approach to managing risk demonstrates respect for dignity, equality, and freedom of older adults (Ibrahim et al., 2019). For older adults who are deemed to lack the decision-making capacity, decisions about their care and best interests are usually guided by advance directives or the substitute-decision makers. However, they should participate in the discussion about risk and their past and present values and preferences must be taken into consideration when planning risk agreements.

6.4.1   BALANCING ACT

Until recently, the focus on assistive technologies has been in controlled environments, such as hospitals and long-term care facilities, where caregivers, persons with dementia, and older adults are the key stakeholders. However, as noted by Niemeijer et al. (2014), in relation to autonomy, societal values must come into consideration if a health care professional permits harm to come to a person who wants to live with risk (Box 6.3). Imagine being a driver on a busy road where you almost collided with a confused woman who ran right in front of your car. From a societal perspective, discussions need to include what risks the wider society is able to make to ensure that the autonomy of older adults and persons living with dementia is improved in public spaces, and in what circumstances values such as safety and harm would win out over autonomy (Burmeister, 2016).

Figure 6.2: Balancing act between risk and autonomy. Courtesy of Hector Perez.

It is challenging to identify what forms of risk are accepted among older adults and persons living with dementia due to the complexities that come with perceived risk. As a society we all live in a world of risk, and it is our ability to be autonomous that allows us to freely make calculated decisions based on these risks. For example, we may cross the street of a busy intersection, knowing the risks of being hit by oncoming traffic if we don't look both ways. We may also make the decision to continue to smoke cigarettes, like Edmund in Chapter 4, despite understanding the heightened risk of worsening his Chronic Obstructive Pulmonary Disease (COPD) symptoms while engaging in this activity. When a vulnerable older adult is being cared by a caregiver, perceived risk shifts to what the decision maker deems acceptable.

There can be multiple conflicting perspectives on the perceived risk of a given activity or behavior among those responsible for a person's care, which in turn leads to a balancing act between their right to be autonomous and their safety. Tensions between persons living with dementia and their caregivers, for example, are all too common when supporting autonomy and minimizing risk (Bunn et al., 2012; Robinson et al., 2007). In a review by Bhatt et al., (2020), in the presence of risk, some caregivers were able to facilitate activities such as driving in the fact of deteriorating ability, upholding the person's freedom of choice: "*[wife] we've discussed this issue about him losing his license eventually because his brother had a stroke and he eventually had to give up his license. So, one of these days it will come to that. And I think if we keep educating him and keep telling him [it will help]*" (Adler, 2010, p. 51).

In other circumstances, however, this notion of risk can lead to decision making occurring outside the person's freedom of choice as the older adult or person living with dementia was excluded from contributing to the outcome. High risk, for example, was found at lower levels of decisional involvement from the person living with dementia, and when a particular conclusion was needed (e.g., the person living with dementia discontinuing driving), it became difficult for a caregiver to stay in a supportive role (Adler, 2010; Fetherstonhaugh et al., 2019; Smebye et al., 2012). As a result, caregivers made decisions based on their own beliefs, overriding those of the person living with dementia, justifying their involvement as for the person's "own good" (Fetherstonhaugh et al., 2019). Therefore, it is imperative that the perspectives of all involved parties, from the care recipient to the care provider, are freely discussed and compromises are made ahead of time to help ensure the autonomy of the older adult or person living with dementia remains as intact as possible.

Technology as an Area of Conflict Between Autonomy and Safety/Risk

While technologies, such as locator devices, could be used to mitigate the risk of harm among older adults and persons living with dementia, this also has the potential to increase risk by providing the end user and caregivers with a false sense of security (Desai and McKinnon, 2020). In the case of wandering behavior in long-term care, for example, the implementation of monitoring technologies could lead to a delayed response from staff due to their reliance on the devices to keep track of residents. In the event that a person with dementia leaves the facility unattended when the devices are malfunctioning, it raises the risk of staff being unaware that the person is missing for a longer period of time (Aud, 2004). As noted by Müller et al. (2017), monitoring technologies are susceptible to errors, malfunctions, and breakdowns. Therefore, despite the claims many technologies provide in terms of reducing the risk of harm caused by wandering, falls, etc., back-up strategies need to be in place if they stop working.

Technologies may also increase the number of risky activities the older adult or persons living with dementia may engage in. In the context of Roger and Michaela's personas, the use of a GPS device may encourage them to go on solo road trips more often or go out for walks in their community. However, this can expose other individuals to other risks while engaging in the activities, such as getting into a car accident or the risk of falling when out on a hike in the mountains. Like the balancing act described in this chapter on varying perceptions of risk among persons living with dementia, older adults, health service providers and caregivers, there is also a balancing act in terms of what potential risks such stakeholders are willing to accept when involving technologies to reduce the risk of harm.

FIND OUT MORE

1. Finding Your Way® Program's Living Safely with Dementia Resource Guide: http://findingyourwayontario.ca/wp-content/uploads/2017/11/Finding_Your_Way_Living_Safely_with_Dementia_Resource_Guide_EN_Final.pdf

2. Let's Talk about Dementia: https://www.paho.org/lets-talk-about-dementia/warning-signs/

CHAPTER 7

Privacy

WHAT IS IN THIS CHAPTER?

This chapter describes the role of privacy on autonomy among older adults. Following this, ethical tensions between privacy, technologies, and autonomy will be described and how it can be seen as a double-edged sword where technology can be used to enable older adults to live the way they want, but at the same time has the potential to compromise privacy.

7.1   ROLE OF PRIVACY ON AUTONOMY

With a globally aging population, there has been increased attention on the role of technology to support healthy aging (Bennett, 2019). While monitoring technologies, such as those that will be described in Part III of this book, provide promise in improving the independence of older adults, they have the potential to reduce their autonomy by treating them as passive subjects whose privacy can be disregarded (McNeill et al., 2017). **Privacy**, as described by Westin (1967), is the freedom to disclose (or not disclose) personal information without incurring unwanted social control by others. Privacy has historically been conceptualized as a static term as only involving intimate actions such as toileting and bathing tasks; however, it also enables a person to engage in the creation of "self" and has the capability to fulfill a variety of psychological needs, such as intimacy, anonymity, and solitude (McNeill et al., 2017). From the perspective of older adults, many have grown up with the idea of privacy as a valuable human right, where information sharing has its rules and limits. Many of these rules, however, are defined by social norms that have been rapidly changing, particularly among generations that have been brought up during the digital age. As a result, today's older adults may be more averse to the sharing of personal information in comparison to the generations that follow (Frik et al., 2019).

> Box 7.1: The seven functions of privacy
>
> Privacy is said to serve seven functions: self-protection, autonomy, emotional release, confiding, social identity, self-concept, and protecting others (McNeill et al., 2017).
>
> - *Self-protection*: The function of self-protection in privacy is to protect from the disclosure of harmful or sensitive information: older adults want to protect their information to avoid verbal conflict, insults, judgment, and the possibility of being taken advantage of in a vulnerable state.

- *Autonomy*: The function of autonomy in privacy entails not sharing data with others against one's wishes: it is necessary to engage users in how data is collected and disseminated.
- *Emotional release*: The function of emotional release in privacy refers to the need to relax from social pressures without fear of others looking on: older adults want freedom from being observed from others.
- *Confiding*: The function of confiding is the ability to control who gets to know information and when: this includes sharing the amount of information one is comfortable with.
- *Social identity*: The function of social identity is about maintaining an image of the self that the individual wishes to convey to others: there is a desire not to share embarrassing information such as physical health concerns.
- *Self-concept:* The function of self-concept involves technology revealing details about the individual's health that they do not know: this new-found information has the potential to harm self-concept.
- *Protecting others*: The function of protecting others in privacy is a desire to withhold information from family or friends so as not to cause alarm.

Privacy and autonomy are related to one's ability to maintain control over their individual circumstance. Within the context of long-term care facilities, for example, respecting residents' personal space has been found to play a major role in maintaining privacy and therefore their dignity (Hall et al., 2014). In addition, the act of seeking consent to attain and share personal information, and subsequently respecting their wishes has also been highlighted as an essential connection between the two factors (Hall et al., 2014). Technology can be a threat to the privacy of personal information. This is evident in consumer-driven societies that convert personal information to commodities without permission from end users. In 2015, Facebook gave unauthorized and unfettered access to personally identifiable information of more than 87 million users to the data firm Cambridge Analytica (Davies, 2015). This has since exposed the hidden use of second-party data among companies and governments around the world (Schneider et al., 2017). Whether the greatest privacy risks come from government commercial entities, or even internet hackers, advances in data collection and processing have the potential to impair our ability to control how much, when, and for how long our personal information is accessible to others (Shapiro and Baker, 2001). Invasion of privacy and the inherent risk associated with the use of technologies that collect personal information, however, are not limited to just third-party sources. As described by Solove's taxonomy (Frik et al., 2019), there are four types of harmful activities associated with data and privacy.

1. Information collection

   a. The lack of transparency about information gathering, lack of effective consent mechanisms, and transparency regarding data collection practices

   b. The inability to control passive audio and video collection by phones and computers, especially in emerging technologies like smart TVs, fall detectors, voice assistants, and home-control systems

2. Information processing

   a. Fears about information disclosure limit willingness to engage in online political discourse

   b. Concern medical staff may misuse data (e.g., assign unnecessary or more expensive treatments)

   c. Potential for fraud on dating sites

3. Information dissemination

   a. Older adults were primarily concerned with their personal information being sold for profit, being disclosed with malicious intent to cause reputational damage, humiliation, or embarrassment

4. Privacy invasion

   a. Participants were particularly concerned about location data and data about their in-home activities, which some saw as sources of compromising information that could facilitate physical attacks on them for their property

Box 7.2: Perception of privacy and technology on older adults

"That's fine, you can take all the data you want, I mean…but is it gonna be of benefit to me?"

-Older adult

(Wang et al., 2019)

7.2   ETHICAL TENSIONS BETWEEN PRIVACY AND AUTONOMY

As highlighted in Chapter 6, there are several ethical tensions as it pertains to permissible risk and autonomy. Similar to this chapter, privacy is also known to have considerable debate about its implications on autonomy. Also known as a double-edged sword, technologies can be used to enable older adults to live the way they want, but at the same time, technologies have the potential to compromise their privacy. For the remainder of this chapter, both sides of the issue in terms of the benefits, consequences, and subsequent recommendations on privacy, technology, and autonomy will be discussed in further detail.

Figure 7.1: Security of mobile applications. Shutterstock: aslysun.

Elements of Technologies that May Enhance the Privacy and Autonomy of Older Adults

Various technologies, from smart homes, locator devices, to fitness trackers now exist to monitor and assist in improving the well-being of older adults. "Smart" technologies for older adults are aimed at sustaining independent living and mitigating health issues via early detection. Thus, monitoring technologies have the potential to provide older adults with a greater sense of choice and control to live the lifestyle of their choosing, whether it be aging in place, or engaging in meaningful activities for a longer period of time.

The evolving nature of decentralized technologies, such as self-sovereign identity that was discussed in Chapter 5, also gives promise for older adults to have more control over the privacy of their data. As noted by Califf and Muhlbaier (2003), more older adults are seeking personal choice and control over their personal information. When older adults have confidential management of their own health data, they experience a sense of autonomy; this has the potential to improve data flow among providers and health care facilities (Califf and Muhlbaier, 2003).

Elements of Technologies that May Compromise the Privacy and Autonomy of Older Adults

Despite older adults becoming primary users of emerging smart systems in health care, some devices take little consideration of the needs of older users, and can pose serious security and privacy concerns (Capgemini Research Institute, 2019; Frik et al., 2019) (Figure 7.1). Due to limited technological literacy and experience, and because of declining mental and physical abilities, many older adults are particularly unaware of the privacy and security risks that are often associated with technologies, such as what information is being compiled about them, and how safe their information is from hackers (Caine et al., 2010). The limited transparency of technologies among some older adults therefore makes this population a larger target for attacks by hackers or scams than younger populations (Caine et al., 2010; Frik et al., 2019) (Figure 7.2).

Figure 7.2: Security of internet sources. Shutterstock: elenabsl.

Common concerns held by older adults about monitoring technologies include invisible audiences, who accesses the data, how often, what level of data and the absence of feedback when systems are in use or when data is accessed (Frik et al., 2019). These concerns can contribute to general feelings of helplessness about maintaining control over the collection and use of their personal data (Frik et al., 2019). There are also fears that the results of the collected data could lead to a subsequent loss of their autonomy, such as being labeled as not having the capacity to make decisions or being reprimanded for engaging in "risky" behaviors and activities (Berridge and Wetle, 2020). While many older adults want to continue to live independently and are aware that technologies, such as home care surveillance, have the potential to prolong their time at home, some worry that they need to balance privacy concerns with the benefits of care surveillance in preserving their autonomy. As highlighted by a participant included in Frik et al. (2019) they state:

> "I would probably choose [a wall sensor that detects] presence over having to share a room with somebody being in a nursing home. So, if I could stay in my own abode [...] that is a concession that I would make." (Frik et al., 2019, p. 25)

Fall detection devices also run the risk of compromising the user's privacy, such as through the use of video recordings. When these devices are used, the viewer's ability to see exactly what the user was doing can be a significant invasion of an older adult's privacy. Its extent may depend on which rooms contain a camera, such as bathrooms and bedrooms. Unfortunately, it is often these same rooms where the risk of falling is most significant (Ganyo et al., 2011). This issue could be further exacerbated if the device was implemented without seeking consent from the older adult, and if the data about third parties are collected by motion-detection devices such as bed occupancy sensors, video cameras, and locator devices. Such recorded information might include the collection of other forms of personal data, such as a person's daily routines, which could put some older adults at a greater risk of being preyed upon (Sánchez et al., 2017). It is important to note, however, that there are many fall detection devices under development, such as what is described by Güttler et al. (2018), that do not rely on the use of videos and image capturing to address these indicated privacy issues. Therefore, we expect to see more commercially available fall detection devices of this nature in the near future.

There are also privacy concerns raised when data collected from monitoring technologies are being used for purposes other than those for which they are designed. For example, allowing commercial companies to acquire the collected data to direct marketing strategies to view products, such as what is seen between Google and Facebook, or governmental agencies that might use the data about a person's functional abilities to make changes to their eligibility for publicly funded supports or benefits payments (Percival et al., 2008).

Recommendations

While older adults often express privacy concerns in relation to technology and autonomy, their views continue to be underrepresented in privacy and security research. Therefore, future research and work in the area of aging and technologies should consider designing systems that empower older adults to make informed decisions, to maintain better security practices, and to have better control over their personal data. In addition, due to many older adults' limited understanding of newer technologies and the data collected, their particular concerns, misconceptions, and blind spots related to privacy and security could be addressed through tailored training and educational efforts (Frik et al., 2019). This could include educational programming, such as the development of security and privacy materials specifically designed for this age group, in addition to materials that address issues of most concern to this population, such as misconceptions about data collection, surveillance, and data sharing. It is important to note, however, that older adults constitute a widely heterogeneous group with different experiences, digital literacy, and socioeconomics, such as Marni, Michaela, Gladys, Roger, and Edmund's personas highlighted in this book. Engagement in social media should also not be overlooked. As well, the potential risks of using hand-me-down or public devices, and how to mitigate them, should be considered.

PART III

How Can Technology Support One's Independence?

Technology is nothing. What's important is that you have faith in people, that they're basically good and smart—and if you give them tools, they'll do wonderful things with them.

Steve Jobs

PIII.1 THE CHALLENGE

Independence, which we define as the degree to which a person can perform a task or activity, is central to notions of healthy aging, successful aging, and active aging. Healthy aging hinges on improving or maintaining ability with advancing age so that an older adult can "be and do what they have reason to value" (World Health Organization, 2015, p. 28). These domains of types of ability include: moving around; building and maintaining relationships; learning, growing, and making decisions; and contributing (World Health Organization, 2015). While decision making is related to autonomy, the other abilities are fundamentally about activities: about doing or participating in something. Importantly, some older adults also identify activity as a contributing factor toward successful aging (Jopp et al., 2015).

As described in Chapter 2, ability is influenced by health conditions (i.e., the presence and severity of diseases, disorders, and injuries) and contextual factors (i.e., personal and environmental factors). AgeTech can enable people to do the tasks and activities that they need to do and want to do by supporting their abilities and reducing barriers in their environments so that they may age well. The challenge, then, becomes designing, developing, deploying, and implementing AgeTech so that some older adults can continue to engage in the tasks and activities of everyday life. The point should not only be to stay alive but to live one's life: to do those things that bring meaning, purpose, and value.

PIII.2 WHAT IS IN PART III?

In this part of the book, we consider how technologies can support independence in key sets of activities relating to older adults looking after themselves (i.e., self care), participating in communities and societies, and enjoying life. These are described in five chapters:

- Self care pertaining to activities that are traditionally known as activities of daily living (ADLs) and instrumental activities of daily living (IADLs)

- Self care activities centered on managing one's health

- Activities for economic and social participation

- Activities for enjoyment and self-fulfillment

- Activities pertaining to mobility in the community.

Each of these chapters begins with a description of the activity and its role in the lives of older adults. We then showcase examples of some of the technologies that can enable some older adults to participate in these activities. In these chapters, we also weave in the real-life stories of the older adults featured in our personas and scenarios. These are followed by a chapter on technologies to support independence in older adults' abilities to move through their communities.

Part III concludes with a chapter that introduces, describes, and discusses the concepts of the acceptability, adoption, and usability of technological and health care interventions. We also explain the factors that facilitate and the barriers that hinder technology acceptance, adoption, and use by older adults. We finish by presenting an example of how to determine the technology acceptance and usability of GPS for persons living with dementia and their caregivers.

PERSONA, SCENARIO, AND SOLUTION

Betty Chung

- **Age:** 76
- **Gender:** Woman, pronouns are she/her
- **Lives in:** Single family home in Winnipeg, Manitoba with 83-year-old husband and adult son
- **Social circle:** Aside from husband and adult son, she has two adult daughters in Winnipeg and one son in Toronto. All three are married and have children. Betty speaks with all of her children daily. Her best friend, Iris, lives within walking distance.

Signature interests
Gardening, swimming, looking after her grandchildren.

" I know that it will be their duty to look after me when I can't look after myself anymore. Right now, it is my duty to look after myself and to look after them."

Health
Type II diabetes, osteoarthritis, osteoporosis, high blood pressure, cataracts, mild cognitive impairment. History of heart attack and knee replacement.

Technology
Has an iPad and an Internet connection. Son has a laptop. Adult children can help with set-up.

Persona. Betty is a 76-year-old woman of Chinese descent who lives with her 83-year-old husband, David, in a single-family home in Winnipeg, Manitoba. Betty emigrated to Canada in the mid-1960s. Betty has Type II diabetes, osteoarthritis, osteoporosis, high blood pressure, and had a heart attack eight years ago. She has a mild cognitive impairment secondary to the heart attack, but this has not affected her day-to-day life or activities. Betty had a knee replacement four years ago and is awaiting cataract surgery. David is in good health but since the COVID-19 pandemic, his mental health has declined.

Betty and David have four adult children. Her oldest son lives in Toronto with his young family. Her two daughters live in Winnipeg with their families, and, prior to the COVID-19 pandemic, they visited at least once a week. Betty's youngest son has mental health challenges and lives with his parents for short periods of time when he needs support or is unemployed. He helps his parents look after their yard, shoveling snow in the winter and cutting grass in the summer.

Looking after her family is important to Betty. She takes pride in ensuring that her family is healthy, happy, and successful. Prior to the pandemic, Betty and David looked after her four young grandchildren part-time while her daughters worked.

Betty used to volunteer as a friendly visitor at a long-term care facility and would like to resume this volunteer position when COVID-19 restrictions are lifted. She is an avid swimmer and enjoys looking after her flower garden.

Betty and David have a landline and internet connection. Betty was given an iPad by her children. However, she doesn't know how to use it. David also has a laptop which he uses occasionally. Betty does not have a smartphone; she feels that they are "a waste of money" since she and David spend much of their time at home and are able to communicate using their landline.

Scenario. Betty relies on David and her neighbor Iris for rides. She is worried that her husband isn't "as sharp" as he used to be; she gets the sense that he would rather stay home than be her driver. Betty is willing to take public transit, but her family is worried about her safety. Not only is Betty somewhat unstable on her feet but the heart attack left her facing some mild cognitive impairments. When she is stressed or feels rushed, she has difficulty navigating unfamiliar areas.

Betty and Iris visited the local swimming pool at least twice a week prior to the COVID-19 pandemic. Since COVID-19 restrictions were implemented, Betty does not see Iris in person, nor does she go to the pool. Yet, Betty states "I feel so much better when I move. The warm water helps soothe my joints. I feel younger and more like myself in the water. I feel badly asking my husband to drive me to the pool. He doesn't like to swim like I do. I need to find a way to keep active".

Betty takes medication for her health conditions and has occasionally missed doses. Her daughter thinks that this has happened more frequently in the last few months. She has also noticed that Betty has misplaced household items more often than usual.

Betty sees herself as the caregiver for her husband, son, daughters, and grandchildren. She does not like to ask for help from them unless necessary.

I don't want to have another heart attack. I know I need to be careful about what I eat and make sure that my blood pressure doesn't get too high. My son in Toronto is also really worried about my blood pressure.

I'm worried about having another heart attack because of the stress that I sometimes feel. I worry about my son and my husband. What will happen to my son when my husband and I are no longer here?

Betty's children gave her an iPad for Christmas. She uses it infrequently because she does not feel confident as she has never received instructions, it is difficult for her to see, and she is concerned about having her identity stolen. She states, "I get lost between the different pages and I can't find the buttons I need to press to get it to work." However, Betty would like to use videoconferencing with her son and grandchildren in Toronto.

Solution. There are many solutions that could support Betty's independence in the activities that matter to her. Given her desire to maintain her strength and activity but to do so without relying on her husband for transportation, she and her family decide to use her existing technologies to help her to set up an exercise plan. Betty prioritizes this goal over the others as she feels that it will help her not only to maintain her health but doing so will also remove demands on her husband. It also reduces the risk of exposure to the COVID-19 virus and will help her to work on her strength and balance.

After clearing the plan with her doctor, Betty's daughter helps her to search for exercise videos that are appropriate for older adults. This low cost and simple solution enables Betty to use the resources that she has: an iPad, internet connection, and the support of her daughter to identify appropriate classes and to use her iPad to watch exercise videos, her son to provide in-person tech support. As she becomes more comfortable with the iPad, her daughter encourages Betty to take online technology classes through public libraries, including those outside Winnipeg, and other community resources (e.g., Connected Canadians). Her daughter also helps her to apply strategies to manage her visual challenges, such as how to enlarge font and use dictation features.

Betty and her children's second priority was to communicate with her family using videoconferencing. FamliNet (2021) was set up on the family's laptop, creating an "online circle" which allows Betty to communicate with her son and daughters. FamliNet gives Betty the ability to initiate conversations with her family at her will and on her terms rather than relying on others to help her. It also allows her son in Toronto to feel more actively involved in supporting her health as it enables family to coordinate care for their loved ones. Unlike other apps, FamliNet was designed with the needs of those with low vision, hearing loss, and fine motor challenges in mind. It can reduce barriers due to language differences by transcribing speech to text, text to speech, and translating conversations. Despite Betty's visual challenges and inexperience using technology, she is able to socialize with her children and grandchildren.

CHAPTER 8

Technology to Facilitate Independence in Self Care—ADL and IADL

WHAT IS IN THIS CHAPTER?

In this chapter we describe independence in self care activities. We focus specifically on activities related to ADLs and IADLs and highlight technologies to support independence in these domains. We then illuminate the potential of smart home technologies to support older adults' independence in ADLs and IADLs.

8.1 WHAT IS SELF CARE?

Older adults need to be able to be independent, or partially independent, in essential activities to live in their own homes in a community setting (Høy et al., 2007; Söderhamn et al., 2013). These essential activities that are performed for the purpose of looking after oneself are collectively known as self care. Whereas some activities in older adults' daily lives are elective or voluntary (i.e., nice-to-dos), self care activities are the must-dos. They are traditionally associated with the purposes of taking care of one's body and promotion of one's personal health (Cook et al., 2020; Laposha and Smallfield, 2020). Self care activities are often classified into basic and instrumental ADLs. **Basic (or personal) ADLs** include care and hygiene tasks such as toileting and management of continence, dressing, grooming, feeding, bathing/showering, and basic mobility in bed and in one's immediate surroundings (Jacobs and Simon, 2015; Matuska and Christiansen, 2011; Polatajko et al., 2007; Whitehead et al., 2013). IADLs, on the other hand, are those that typically involve greater cognitive demands and thus are considered more advanced tasks (Czarnuch and Mihailidis, 2011). Challenges performing IADLs often occur before difficulties with ADLs arise (Hellström and Hallberg, 2004). IADLs include money and medication management, cooking, doing laundry, grocery shopping, looking after one's home (e.g., cleaning, basic maintenance), using the phone, and community mobility (Polatajko et al., 2007; Whitehead et al., 2013). Some older adults identify that self care activities and independence are inextricably linked (Dale et al., 2012; Söderhamn et al., 2013).

There is a need to reconceptualize and expand the notion of self care. In the era of the Fourth Industrial Revolution in which technologies are embedded in our daily lives (Liu, 2018), the way that older adults perform self care activities has changed. Grocery shopping can be done in the comfort of home. Whereas visiting a bank was once an essential aspect of assessing an older adult's ability to live independently in a community setting, we can now pay bills online and are rarely required to visit a bank. We make appointments using online booking systems rather than by telephone, and we no longer need to use paper-based maps and schedules to plan public transit routes. A Roomba robot vacuum now cleans the floor for us. Indeed, how we do self care has changed.

The essence of what self care encompasses has also shifted. The COVID-19 pandemic has expanded how we think about self care and emphasized the importance of exercise, sleep, and other activities that contribute to physical and mental wellness including among older adults (Morrow-Howell et al., 2020). Social relationships, mindfulness, and mentally stimulating activities are also components of self care (Cook-Cottone and Guyker, 2018; Dale et al., 2012; Göransson et al., 2017; Laposha and Smallfield, 2020; Loeb, 2006). In an era of moving toward partnerships between people and their health providers, older adults can now be more involved in managing their own health. There is greater onus on managing one's health behaviors (e.g., diet, exercise, medication regimen), monitoring vital signs (e.g., blood glucose, blood pressure, heart rate), and coordination of health care services (Rogers and Mitzner, 2017). As a result, looking after yourself not only means taking care of one's hygiene and basic needs but also performing activities to maintain health and wellness. These will be examined in Chapter 9 which focuses on self care activities aimed at managing one's health.

8.2   TECHNOLOGIES TO SUPPORT PERFORMANCE OF ADL

A plethora of assistive products exist to help older (but also younger) adults perform ADLs. A stabilizing glove, called Steadi-One (https://steadiwear.com/; Figure 8.1), has been developed for younger and older adults living with Parkinson's disease and essential tremor. This AGE-WELL-supported technology helps to stabilize involuntary movements that are experienced by people with tremors. A preliminary study has demonstrated that the Steadi-One reduces tremors resulting in better hand control (Sampalli et al., 2020). This improves people's ability to eat, drink, write, and groom themselves (e.g., apply make-up, shave). A second-generation stabilizing glove that can adapt to a wider range of hand tremors in people with essential tremor and Parkinson's disease is available for pre-order with a delivery date in the second quarter of 2022.

Figure 8.1: The Steadi-One stabilizing glove was created by Steadiwear Inc. Photo by John Hryniuk. Courtesy of AGE-WELL.

Arthritis is a common condition among older adults that affects independence and quality of life (Anderson and Loeser, 2010; Havens et al., 2017; Hootman et al., 2012). Older adults who have arthritis may use reachers to help them safely pick up objects. People with hand arthritis may have difficulty using standard reachers due to poor grip strength. A lightweight electronic reacher was created for people with arthritis and limited grip strength, resulting in a device that helps them to get the everyday objects needed to perform their ADLs, thereby increasing safety and independence.

The above examples are only two of thousands of relatively low technology commercially available products and can help people with disabilities as well as older adults to perform ADLs. Other products utilize high technologies. Braze Mobility has developed a blind spot sensor system for people who use wheelchairs due to conditions such as stroke, multiple sclerosis, and spinal cord injuries (https://brazemobility.com/; Figure 8.2). Because of limited trunk mobility, some wheelchair users cannot see objects behind them. This affects their ability to maneuver in small spaces and results in getting stuck or damaging walls and furniture. The Braze Mobility system can be attached to conventional wheelchairs to create smart wheelchairs, enabling users to effectively navigate through tight spaces such as bathrooms and hallways. Thus, it provides access to areas needed to perform self care tasks such as toileting, bathing, and cooking.

Figure 8.2: Braze Mobility Inc.'s blind spot sensor systems can be installed on any wheelchair. Courtesy of Braze Mobility.

Advancements have also been made in one of the essential, yet most challenging, ADLs for older adults with physical and cognitive challenges: showering and bathing. Difficulties with showering and bathing is the strongest predictor of relocation to an institutional setting (Fong et al., 2015) and older adults require help to bathe or shower more often than all other ADLs (Wiener et al., 1990). Yet, some older adults are also reluctant to receive help for bathing or showering from paid caregivers and family members because of its highly intimate nature (Ahluwalia et al., 2010). Thus, a solution that supports older adults' independence and also respects privacy and manages safety for staff and older adults is important.

Solutions include Poseidon, a robotic shower developed in Sweden (http://roboticscare.com/poseidon/). Unlike other bathing systems, the hollow shower chair moves toward the user rather than the user moving forward into the shower. After being seated, a mechanical arm transfers the user into the shower and the shower door closes. Several moveable showerheads, controlled by the user, emit soap and water. Findings from a small study of older adults in an assisted living facility found that using the robotic shower was empowering and partially increased their independence, even though some still required assistance with undressing/dressing, drying off, and transferring to the shower chair (Bäccman et al., 2020). Involvement in the process of showering provided the older adults a feeling of control despite residing in an assisted living setting. Other innovations in showering are also being developed, such as I-Support, an assisted bathing platform designed specifically for the older adult population. The I-Support system contains several technologies including activity monitoring and recognition, a motorized transfer chair, a series of sensors that perceive the user's movements as well as environmental conditions in the shower (e.g., temperature, humidity, water flow), and a soft robotic arm that assists the user with washing tasks (Zlatintsi et al., 2020). The I-Support system has been the subject of two validation studies which reported that I-Support was useful for bathing tasks and was highly acceptable among older adults.

The technologies described above provide support to older adults with physical limitations so that they may perform ADLs. Technologies to assist older adults with cognitive challenges also exist. Devices, apps, virtual coaches, and robots can provide step-by-step instructions and reminders to help older adults with cognitive challenges to dress themselves (Bewernitz et al., 2009; Burleson et al., 2018) brush their teeth (Bewernitz et al., 2009), drink water (Bewernitz et al., 2009), and wash their hands (Mihailidis et al., 2008). While these technologies are promising, especially robots, their commercial availability is limited. Most have been developed and tested as prototypes and not all have been thoroughly studied (Shishehgar et al., 2018). In comparison to other conditions, there appear to be fewer technology solutions for persons living with dementia focused specifically on improving performance of ADLs (Patomella et al., 2018; Van der Roest et al., 2017). Evaluation in real-world and care situations with persons living with dementia is also needed (Van Aerschot and Parviainen, 2020).

8.3   TECHNOLOGIES TO SUPPORT PERFORMANCE OF IADL

In the following section, examples of technologies to support independence in IADLs, specifically medication management, meal preparation, grocery shopping, and financial management are presented. These represent only several of the numerous assistive products, everyday digital technologies, and high tech devices available to older adults.

Older adults' ability to manage (i.e., taking medication on time as prescribed) their medications is an essential IADL. People aged 50 years and older comprise the majority of medication users (Ramage-Morin, 2009) and over 90% of older adults take at least one medication per day (Qato et al., 2008). Cognitive impairment and dementia impact older adults' ability to manage their medications, especially when medication regimens are complex (Campbell et al., 2012), and problems managing medications can precipitate moving to assisted living (Mitzner et al., 2014).

Mobile health (also known as mHealth) applications can help older (as well as younger) people to adhere to their regimens by providing reminders, obtaining refills, recording doses taken, and storing medication information (Park et al., 2019). Hundreds of medication management apps are available through Google Play and the iOS Apple App Store (Tabi et al., 2019). Examples include MyMedRec (https://www.knowledgeisthebestmedicine.org/index.php/en/app/), MediSafe (https://www.medisafeapp.com/), Dosecast (http://www.montunosoftware.com/about/), and Med Helper (https://medhelper.com/) and can be used in combination with low tech pill boxes (Figure 8.3). Such apps can also help older adults better understand, and thus actively manage, their medications (Tabi et al., 2019). While such medication management apps hold promise, they must be usable by older adults and be easy to see and navigate (Grindrod et al., 2014; Stuck et al., 2017). Moreover, apps do not replace reminders from other people (Campbell et al., 2012).

Figure 8.3: mHealth medication management app. Shutterstock: Mix Tape.

Other innovations that not only provide reminders but also dispense medication are available to support independence. MedMinder (https://www.medminder.com/) allows automatic medication dispensing combined with reminders. MedMinder is equipped with wireless technology that does not require an external internet connection. It supports remote caregiving by sending dosage activity and real-time notifications to family members by email or text messages. Smart pillboxes (i.e., pill/medication dispensers) are also available that automatically release the correct dose to reduce the risk of adverse effects of taking an incorrect dose (e.g., PilloHealth, 2019; Pivotell, 2016; Tricella, 2020). Some of these devices can improve medication adherence among persons living with mild cognitive impairment (Kamimura et al., 2012). Robots offer another possibility for providing medication reminders (Shishehgar et al., 2018). For Betty, featured in the persona at the beginning of this section, a simple medication reminder system may help her to better manage her medication without having to rely on her adult son or her husband. It could also provide reassurance to her children, confirming that their mother takes her medication on time and as prescribed.

Being able to cook for oneself is an important aspect for aging in place in one's own home. People's dietary behavior and meal preparation habits can be influenced by age-related changes secondary to declining physical function due to chronic conditions (Bostic and McClain, 2017; Whitelock and Ensaff, 2018). Safety during meal preparation tasks can also be a concern, particularly among persons living with cognitive impairment (Ibrahim and Davies, 2012; Kivimäki et al.,

2020). Some older adults are at risk of malnutrition which can result in further health challenges including decreased strength, immune response, and wound healing (Harris et al., 2019; Payne et al., 2020). Yet, maintaining some independence in meal preparation and choice of food is important to some (but not all) older adults, even if they receive support for meal preparation activities from family members or community organizations (e.g., Meals on Wheels) or prefer to utilize take-out services (Whitelock and Ensaff, 2018).

Technologies can facilitate older adults' meal preparation activities (Figure 8.4). The Cook-Stop Automatic Oven Control (https://www.alzstore.com/automatic-oven-control-p/0352.htm) uses motion detection to automatically turn off the stove after the kitchen has been empty for a selected number of minutes (e.g., 5, 10, or 15 minutes). FireAvert (https://www.alzstore.com/fireavert-stove-fire-prevention-p/0106.htm) is a simple device that is plugged into the back of the stove. When smoke is detected, the power to the stove is cut reducing the chance of a fire. A Swedish study explored the experiences of older adults with memory impairment or dementia and their families in response to implementing a stove timer into their routines (Starkhammar and Nygård, 2008). Both groups expressed a greater sense of safety and less worry when the device was installed. For Roger, the person featured in the preceding section who is living well with dementia and who is an avid baker, utilizing a high or low tech device to remind him that the oven is on could help him maintain his independence.

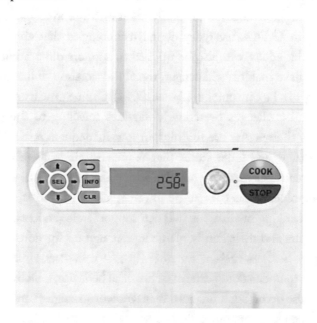

Figure 8.4: CookStop Automatic Oven Control device. Used with permission.

Technologies can provide step-by-step instructions and coaching for food preparation activities. Verbal instructions that are stored on a tablet can be transmitted using a wireless Bluetooth speaker and combined with a talking alarm clock to remind users of meal preparation activities to prepare meals at certain times of day (Lancioni et al., 2012, 2017; Perilli et al., 2013). Digital home assistants (e.g., Google Home, Amazon Echo, and Alexa) are another alternative (Czaja et al., 2019). However, these approaches that employ everyday digital technologies have yet to be extensively tested in real-life settings. Such strategies do not address the issue of safety during meal preparation activities. High tech solutions may provide this advantage; they are designed not only to provide instructions and coaching but also monitor the user by employing sensors. Chore or domestic robots can support performance in IADLs such as cooking and cleaning (Shishehgar et al., 2018). Smart kitchens and appliances have also been developed, designed, and tested with persons with early-stage dementia (Menghi et al., 2018). Their smart kitchen platform not only provided instructional support for meal preparation but also linked kitchen appliances and their multiple sensors. Bouchard et al (2020) created a smart range that provides guidance and monitoring for people with cognitive impairment as they prepare meals. Sensors estimate cooking time and can detect fires. The smart range also provides audible and visual alerts if safety concerns are recognized and can also shut itself down. However, these high tech devices are in prototype form and lack extensive testing in community situations with end users.

The preceding paragraphs focused primarily on assistive products and high tech devices to support older adults with IADLs. Everyday digital technologies that are available to the larger population, not only older adults, can also be utilized to support older adults' independence. For those older adults who have mobility challenges, the ability to shop online and receive their purchases by home delivery can be convenient (Lignell, 2014). Online grocery shopping can help older adults who have physical challenges overcome difficulties traveling to the store, and also avoid in-store barriers such as shelves that are too high or low, inadequate rest and bathroom facilities, narrow aisles, poor signage, and large stores that require significant walking (Lesakova, 2016). Online shopping has become increasingly popular during the COVID-19 pandemic (Aston et al., 2020), including among older adults (Morrow-Howell et al., 2020; Soh et al., 2020). Solutions are available through most banking institutions that can support independence and safety in financial management. Older adults and their family members can sign up for notifications in the event of unusual banking activity. Services such as Eversafe (https://www.eversafe.com/) are available that use artificial intelligence to understand patterns of financial behavior, look for irregularities, and notify subscribers when these occur. Such financial management strategies can facilitate independence among older adults while also providing a desired safety net, especially given that older adults are at increased risk of frauds and scams (Optimal Aging Portal Blog Team, 2018; Simhon and Trites, 2017). However, some people may still prefer traditional in-person banking or may seek assistance from family members to help them manage their money such as Gladys, one of the older adults featured in Chapter 2.

8.4   SMART HOMES TO SUPPORT PERFORMANCE OF ADL AND IADL

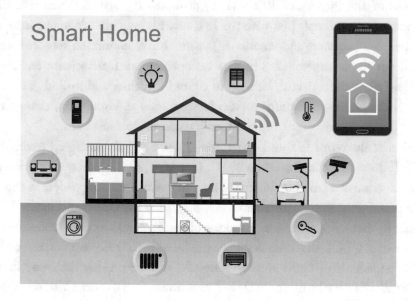

Figure 8.5: Smart home.

The term smart home (Figure 8.5) is becoming an increasingly common term in our everyday language. But what does the term smart home mean? The meaning is variable and is dependent on a smart home's technological aspects and usage (Gutman et al., 2016). In the more broad sense, a smart home has been defined as consisting of "networked devices and related services that enable home automation for private end users" (Statista, n.d.). In this definition, a device has the capability to be connected directly or indirectly via a so-called gateway to the Internet. In contrast, Liu et al. (2016, p. 45) defined smart homes as "a special kind of home or residence equipped with sensors and actuators, integrated into the infrastructure of the residence, intended to monitor the context of the inhabitant to improve a person's experience at home." Overall, smart homes can be divided into six segments: control and connectivity (smart speakers, personal voice assistants); smart appliances (fridges, washing machines); security (motion sensors, door locks); home entertainment (entertainment remotes); comfort and lighting (smart lighting/bulbs); and energy management (thermostats, radiator controls). In health care, smart homes have been called smart medical homes (Muse et al., 2017) or home health monitoring technologies, defined as an entity intended for "monitoring an occupant's health status to maintain a person's well-being" (Liu et al., 2016, p. 45).

In terms of hardware and software architecture, smart homes can be understood as an entity that has at least four main layers: the acquisition layer (where physical sensors collect information); service layer (where communication occurs between sensors and wired or wireless protocols); application layer (where a graphical interface make data collected from sensors visible to users); and the storage layer (where data are either locally or remotely stored in a database) (Liu et al., 2016).

The arrival of The Internet of Things (IoT) disrupted the way in which smart homes have been traditionally viewed. Simply put, with the IoT, smart homes are becoming smarter and more perceptive to people's behaviors and emotions. The number of connected sensors increases in the homes which leads to massive volumes of data being collected. Data science techniques such as machine learning allow these data to be used to control actuators and to find patterns in human behaviors. Thus, the options to automate everyday processes at home (e.g., change temperature, open and close doors, turn off oven) and to identify patterns is almost infinite.

The fusion of the terms IoT and smart homes has led to an emerging term: The Home of Things (HoT) (Future of Today Institute, 2021). In a HoT, data sensors are not necessarily embedded in the structure of a building. Instead, all the things surrounding people in their homes can be used as sensors. Personal voice assistants (Box 8.1), such as Alexa and Google Home, can be used to connect to thousands of devices. In terms of security and risk mitigation, HoT is starting to show its potential. CLOi autonomous robots, produced by LG Electronics, use ultraviolet light to disinfect high-touch, high-traffic areas in hospitals, and are making their way into our offices and homes (Future of Today Institute, 2021; LG Newsroom, 2020). Smart kitchens that connect to other smart devices are now available in the market, saving users time, money, and energy as well as diminishing risks to a user's safety. Bosch has created a smart kitchen that connects to Nest Protect. If a user forgets about a pot on the stove and it starts to catch fire, the Nest smoke detector will tell the oven to turn itself off (Future of Today Institute, 2021). Thus, the HoT has immense possibilities as it relates to managing risks in the homes of older adults and enhancing their independence, particularly for people living with dementia. In the future, it is expected that the market will be flooded with HoT products. In fact, some already exist. AmazonBasics now makes appliances (coffee maker, instant pot, meat thermometer, microwave) that respond to Alexa voice commands (Future of Today Institute, 2021).

Smart homes permit the collection of information about people's activities through various sensors and ambient devices. These activities are movements and tasks within ADLs and IADLs. This allows determining a baseline of patterns of activity. Having a baseline allows for knowing when something is out of the ordinary. Out of the ordinary may mean that an incident occurred (such as a fall) or that something was missed (i.e., taking medication). When such variations in regular activity patterns occur, notifications can be sent to devices and these can alert older adults, their caregivers with whom they live or those who live outside the home, or health care providers to act. For instance, if an older adult is typically active by 7:00 AM each day and smart home sensors do not detect activity by this time, a device may ask the person whether he or she is alright or send a message to caregivers that no activity has been detected. Caregivers can then take action to call the older adult, physically check in on them, or if the older adult lives in a congregate setting, have a staff member check on them. Similarly, if smart home sensors know that an older adult typically takes medication at a certain time of day, if the system recognizes that the person has not taken their medication, it can remind the person to do so. In this way, smart homes can help older adults live independently but with a safety net in place, providing reassurance to the older adult and peace of mind to the caregiver.

> **Box 8.1: Smart assistants to support independence in self care. Author: Hector Perez (Post-doctoral fellow, University of Waterloo).**
>
> Smart assistants hold enormous benefits for older adults. Because they are voice enabled, they can be used to call emergency services (Rawes, 2021). Also, for individuals with physical limitations, a smart assistant setup can provide the ability to control their home environment from their wheelchairs (Simpson and Levine, 2002) or beds (Fadillah and Ihsan, 2020). As such, they can be helpful for people with mobility issues (Kadylak et al., 2020), vision loss (Sankalpani et al., 2018), and reduced dexterity (Mitroff and Price, 2020). They can also guide people to obtain information about their communities (e.g., locate grocery stores, check on hours of operation of services) and provide instructions. Smart assistants can help older adults age in place (Choi et al., 2018; O'Brien et al., 2020), live safely, and live independently (Ennis et al., 2017).
>
> However, barriers to adopting smart assistants include accuracy, accent, or dialect-related recognition issues, cost, language coverage, end-user expectations, flexibility to meet the user case, operational support, and data and security (Speechmatics, 2020). Also, the information given by a smart assistant must be scrutinized for its accuracy and reliability, particularly when it is related to health care information (Shalini et al., 2019).

Nobi (https://nobi.life/) is an example of a commercially available device that comes in the form of a smart lamp and provides continuous monitoring of older adults at risk of falls. When an older adult gets out of bed in the middle of the night, Nobi recognizes the movement and the light turns on to help the person orient themselves and avoid tripping and subsequent falls. If it detects unusual activity, it speaks to users and requests verbal acknowledgment that they are safe. If there is no response, it can call emergency services and unlock the doors in advance of their arrival. Because Nobi recognizes regular pattern of activities, it can provide cues to the person to complete basic ADL and IADL activities as well as exercise. Other commercially available smart home solutions are available. Grand Care (https://www.grandcare.com/family-caregiving/) combines sensors placed throughout the home with a large touch-screen display. Like Nobi, Grand

Care recognizes activities and provides alerts to the older adult and their caregiver when variations occur. It also provides visual reminders such as a customized activity checklist for the older adult to follow throughout the day as well as vital sign monitoring. Unlike Nobi, Grand Care facilitates communication and social exchanges between older adults and their remote caregivers through videoconferencing, sharing photos, and playing games together.

LISA is a furniture kit that combines sensors and sensor systems, mechanical elements, and communication interfaces to support older adults in their homes (Figure 8.6). Such kits can be customized and installed in private homes as well as care facilities. These strive toward the seamless integration of assistive products and sensors into the built environment. LISA is the result of a European collaboration and its furniture products have been successfully commercialized.

Figure 8.6: LISA. © LISA (Living Independently in Südtirol Alto Adige; T. Linner, J. Güttler, T. Bock in collaboration with MM Design, Pfeifer Planung, GR Research, Kofler Alois, Elektro Haller, TUM), used with permission.

AGE-WELL has supported technology development and research in the area of smart homes, including teams at the University of Alberta led by Eleni Stroulia and at TAFETA Smart Systems for Health (Bruyère Research Institute and Carleton University) led by Frank Knoefel and Rafik Goubran. The second team developed a Night-time Wandering Detection and Diversion system that responds to the needs of caregivers who reside with persons living with dementia (Figure 8.7). The system, composed of pressure mats, sensors, smart bulbs, and speakers, recognizes when the person living with dementia leaves the bed (Ault et al., 2020). It automatically turns on lights and gives audio instructions to help the person find their way to the bathroom and back to bed. If the person with dementia opens a door through which they can exit the home, the caregiver is notified. Otherwise, the caregiver is not disturbed, creating conditions for them to sleep through the night. Findings of a pilot study suggest that this system successfully redirected persons living with dementia when they awoke at night and improved caregivers' depression and anxiety which allowed them to rest more peacefully at night (Ault et al., 2020).

Figure 8.7: Researchers in an "apartment lab" at the AGE-WELL National Innovation Hub called SAM3 test heat-sensing technology that can show the complexity of meals. Other smart-home technologies being developed at the hub include a wandering detection and diversion system, and pressure-sensitive mats to monitor an older adult's health during sleep. Photo by John Hryniuk. Courtesy of AGE-WELL.

FIND OUT MORE

1. Aging and Health Technology Watch: Industry Market Tends, Research & Analysis: https://www.ageinplacetech.com

2. LISA A New Living Concept for the Elderly: https://www.mmdesign.eu/en/design-projects/45-lisa

CHAPTER 9

Technology to Facilitate Independence in Self Care—Health Management

WHAT IS IN THIS CHAPTER?

Unlike older adults in generations before them, today's older adults may now be more responsible for managing their health. Whereas in the past older adults were seen as the recipients of health services, some are now seen as consumers of such services. Some visit their health providers having already "read up" on a condition, explored treatment options, and have a long list of questions to discuss. Some may be more involved in the decision making pertaining to their health and intervention plan and may be actively involved in improving their health by making certain behavioral and lifestyle changes. We view these actions to manage one's health as a component of self care.

In this chapter, we describe how self care activities that center on managing one's health can be supported using technologies. These include technologies focused on access and use of health information, technologies to manage health services, technologies for physiological and activity monitoring, and technologies to support exercise.

9.1 TECHNOLOGIES TO ACCESS AND USE HEALTH INFORMATION

One aspect of managing health is being digitally literate. Digital literacy in health is related to being able to find and analyze health information using electronic sources (Norman and Skinner, 2006). It can include being able to use everyday digital technologies including the Internet, and a computer, tablet, or smartphone to look for information about health conditions, behavioral strategies, and programs (Figure 9.1).

Older Canadian adults represent the fastest growing segment of internet users than any other age group and 88% of those aged 65 years and older report using the Internet daily (AGE-WELL, 2021). However, internet usage is still substantially below younger cohorts (Davidson and Schimmele, 2019). Searching for health information online is one of the most popular activities among adults as well as older adults (Hunsaker and Hargittai, 2018; Jones and Fox, 2009). Online health resources can provide older adults with information as well as allow them to connect with others who have similar health concerns and with their health service providers (Watkins and Xie, 2014). Some older adults indicate that being knowledgeable about health conditions and seeking information to manage their health concerns is an important aspect of self care and, consequently, independence (Göransson et al., 2017). Using the Internet to seek health information and applying it to keep safe has become particularly important to older adults during the COVID-19 pandemic (Chen et al., 2021; von Humboldt et al., 2020).

Figure 9.1: Burn Evans using everyday digital technology. Used with permission.

However, not all older adults have the technology, financial means, or skills to be able to access health information online. Older adults' internet use to access health information is related to education and income (Hargittai et al., 2019; Hunsaker and Hargittai, 2018); those with lower educational attainment and income are less likely to use the Internet for several reasons including the disposable income to purchase devices and an internet connection (Chang et al., 2015). Low vision, cognitive challenges, fine motor, and other disabilities may affect an older adult's ability to use everyday digital technologies and thus go online (Choi and DiNitto, 2013b; Echt and Burridge, 2011; Gell et al., 2015; Gitlow, 2014; Hunsaker and Hargittai, 2018; Kottorp et al., 2016). Concerns related to online safety and privacy may affect older adults' internet use (Quan-Haase and Ho, 2020). These factors contribute to the digital divide between and within cohorts (Hunsaker and Hargittai, 2018).

University of Toronto's Technologies for Aging Gracefully Lab (TAGlab) has been supported by AGE-WELL to develop innovations to support older adults' access to and use of health information online. They are developing a language customization tool to simplify health information that is found online, making health information accessible to those older adults with low literacy (TAGlab, n.d.-a). This team is creating tools to support older adults' use of security technologies (e.g., email filters) so that they may be increasingly aware and proactive about managing their online security (TAGlab, n.d.-b). They have also developed TAGhelper, an interactive aid that helps older adults navigate and use tablets (Conte, 2019; Conte and Munteanu, 2018). These innovations may, in turn, help some older adults to use the Internet to obtain health information and manage their health. These technologies could be very helpful to some older adults who are new tablet and internet users, such as Betty.

9.2   TECHNOLOGIES TO MANAGE HEALTH SERVICES

Everyday digital technologies coupled with internet access also facilitate health management beyond information-seeking. It allows people to communicate with health care providers to schedule appointments, consult about non-urgent issues, request prescription refills, review diagnostic tests, and share information about their symptoms. Interactive platforms can enable older adults to seek information as well as manage specific health conditions such as diabetes and heart conditions (Buck et al., 2017; Ferreira et al., 2019). In many Canadian provinces, governments rely heavily on people's ability and access to everyday digital technologies to trace and schedule COVID-19 tests and vaccinations. Those without such access and skills are left to endure wait times when scheduling these appointments by phone or to rely on family, friends, and community organizations (e.g., cultural associations, libraries) to make these arrangements. COVID-19 test results are shared by email, text message, and automated voice messaging, leaving those without mobile phones, computers, tablets, and an internet connection with fewer notification options.

The COVID-19 pandemic has inspired change in communication practices between health professionals and people. A survey conducted by AGE-WELL in July 2020 highlighted that 52% of Canadians aged 50 years and older had an appointment with a health care provider using telehealth, although only 7% had an appointment by video (AGE-WELL, 2021). Specialized platforms that enable virtual care from physicians and nurse practitioners have been developed, such as Maple (https://www.getmaple.ca/for-you-family/how-it-works/). Virtual care platforms allow some older adults to obtain virtual services and assessments from health care providers online using video, audio, and text messaging. This technology offers an accessible option for those older adults who have non-urgent health concerns but may not be able to visit their general practitioners due to timing (i.e., after hours), mobility or transportation problems, or public health concerns.

While there is evidence that some older adults are willing to use everyday technologies to manage their health (Chen et al., 2021), health care providers cannot assume that all will have sufficient resources and skills to do so (Gordon and Hornbrook, 2018). There are older adults who do not have these prerequisites or who wish to use conventional approaches. Other barriers include language, disability, and limited financial resources. Some older adults need instructions on how to use telehealth services (Gavett et al., 2017; Xie et al., 2012, 2020).

Technology can also support some younger and older adults to manage their health information and increase their involvement in the health services that they receive. An hStick (also known as health stick), contained on a simple USB stick, can store data such as personal information, blood type, health conditions, medications, illnesses, living will, baseline blood pressure, and lifestyle/behavioral information (e.g., exercise, nutrition, sleep) (Pekkarinen et al., 2013), helping those with complex medical problems, chronic conditions, and memory challenges to share information. It may encourage older adults to monitor their health and health behaviors, ultimately enhancing self care and supporting independence.

9.3   TECHNOLOGIES FOR PHYSIOLOGICAL AND ACTIVITY MONITORING

Wearable technologies monitor vital signs (e.g., heart rate, blood pressure, breathing rate) as well as behavior and activity patterns (e.g., physical activity such as number of steps and calories burned, hours slept) (Farivar et al., 2020). They are widely available and include the Fitbit, Apple Watch, Jawbone UP, and many others (Alharbi et al., 2019). Because older adults typically have at least one chronic condition, wearables can be used not only for monitoring purposes but also for older adults to better understand physiological and activity patterns, take action to improve their behaviors, and thus maintain a healthy lifestyle (Farivar et al., 2020). Such wearables can encourage physical activity by sending motivational messages and allowing people to compare their activity from one day to the next and to their peers (Finkelstein et al., 2016; Patel et al., 2015). Some wearables allow the user to set and receive medication and appointment reminders (Farivar et al., 2020). However, some evidence suggests that some wearable technologies worn on the wrist may be less accurate when used by people who have mobility challenges and use devices such as walkers and canes or walk at a slow pace (Floegel et al., 2017; Wong et al., 2018).

VitalTracer (https://vitaltracer.com/home-care-monitoring/; Figure 9.2) is a medical-grade wearable device that provides continuous monitoring of all vital signs, facilitating remote monitoring of patients in health or home care settings (VitalTracer, n.d.). It was recently deployed to detect early COVID-19 symptoms, allowing health care providers to take action to prevent transmission. It has also been used to monitor those recovering from COVID-19.

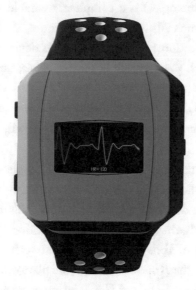

Figure 9.2: VitalTracer™ smart watch. Courtesy of VitalTracer.

Despite the potential of wearable devices to support older adults' independence in self care and some older adults' willingness to use them (Preusse et al., 2017), people aged 65 years and older make up a small minority of users (Farivar et al., 2020). Evidence about long-term adoption of wearables by older adults is conflicting even though enthusiasm to use wearables may initially be high (Brickwood et al., 2020; Lee et al., 2020). Some older adults have concerns about reading and interpreting the outputs of wearables (Farivar et al., 2020). Therefore, training and support can alleviate concerns of older adults who have limited (or different) experience with technology (Rogers et al., 2020). Support from health professionals for older adults to use wearables would facilitate long-term changes in activity levels of seniors (Brickwood et al., 2020). This highlights that technology alone cannot replace social support.

Medical and non-medical wearable devices are becoming another piece of the smart home and HoT ecosystem and bringing with them the ability for live monitoring. The COVID-19 pandemic has accelerated the production and adoption of digital transformation in health care (Future of Today Institute, 2021). New discoveries in diagnostic testing and remote monitoring, supported by cloud computing and machine learning, are disrupting traditional doctor visits. Called doctorless exams, some products are already in the market such as the VROR system that sends data to a mobile app for optic nerve testing (Future of Today Institute, 2021). Similarly, technologies to support at-home lab tests are being developed. Healthy.io's Velieve sells a urinary tract infection test kit that uses a mobile app to connect patients who have positive results with an online doctor. The app also sends a prescription to a nearby pharmacy as needed (Future of Today Institute, 2021). The Zio patch (from iRhythm Technologies) is a complete, wearable, ambulatory cardiac monitoring solution that can be used at home (Future of Today Institute, 2021). Cloud-based Wireless Body Area Networks allow communication from users' wearable devices back to their home computer. Data collected from sensors, such as heart rate or oxygen levels, can be constantly monitored from home. It is expected that the 5G networks will accelerate this new wave of smart device innovations (Future of Today Institute, 2021). Moreover, the shift in the way assessments are being conducted could bring a new approach to current assessments that rely heavily on self-report data and are thus prone to bias.

HoT can bring good things but bad ones as well. People are buying and installing HoT devices without understanding how these devices really work. Software upgrades are left to end users and some of them simply do not have the technical skills nor the technical knowledge to monitor their systems. As a result, they are vulnerable to hacking. The HoT platform providers are not always interoperable (Future of Today Institute, 2021). In other words, Google speakers cannot interact with devices running on the Amazon smart home platform when no interoperability exists. Moreover, there is a lot of hype in the topic of HoT and smart homes and home health monitoring technologies for older adults. Our recent systematic literature review found that the technology readiness level for smart homes and home health monitoring technologies is low; there is a high level of evidence that smart homes and home health monitoring technologies are effective to monitor activities of daily living, cognitive decline and mental health, and heart conditions in older adults with complex needs. There is no evidence that smart homes and home health monitoring technologies help address disability prediction and health-related quality of life or fall prevention (Liu et al., 2016). These findings suggest there is still room for development and research in the area of smart homes and HoT in health.

9.4   TECHNOLOGIES TO SUPPORT EXERCISE

Some older adults perceive physical activities as enjoyable, others do not. For this reason, we include older adults' activities centered on exercise and physical activity in two distinct chapters: the present chapter on self care, and a separate chapter (Chapter 11) on activities for enjoyment and self-fulfillment. While the activities themselves may be similar, the underlying purposes of, or motivations for, these activities are distinct.

Exercise is good for people regardless of age. A physically active lifestyle can counteract or delay age-related decreases in functional capacity as well as reduce risk of disability, morbidity due to various conditions (e.g., cancer, fractures, recurrent falls, depression, dementia), and mortality (Cunningham et al., 2020). Conversely, being physically inactive is a risk factor for chronic conditions and obesity (Cunningham et al., 2020). Aerobic exercise can also benefit older adults who have cognitive impairment and dementia (Nuzum et al., 2020), those who are frail (De Labra et al., 2015), as well as older adults who experience negative mental health due to social isolation resulting from the COVID-19 pandemic (Callow et al., 2020). Given its health-promoting effects, it is suggested that older adults engage in at least 150 minutes of moderate physical activity per week (Canadian Society for Exercise Physiology, 2020; Piercy et al., 2018). While there is evidence to suggest that many of older adults' favorite activities involve physical activity (Szanton et al., 2015), many older adults, like their younger counterparts, do not get as much physical activity as they should (Gomes et al., 2017; Kohl et al., 2012). This is of concern given that exercise is an effective way for older adults to care for themselves.

Technologies to support older adults' exercise are abundant. As described in the preceding section, activity trackers, which have the largest market share of all wearable devices (Mück et al., 2019), can encourage older adults to become aware of their physical activity patterns and increase their activity accordingly. In an AGE-WELL survey completed in 2020, 13% of Canadians aged 65 years and older use activity trackers or apps and 19% of these increased their use in the last year (AGE-WELL, 2021). Roger Marple, the person featured in the preceding section who is living well with dementia, uses an activity tracker to monitor his activity level and to remind him to exercise daily.

Other everyday digital technologies such as computers, tablets, cell phones combined with internet access or applications can also help older adults exercise by providing access to virtual exercise classes specifically for older adults. For older adults who are already physically active, switching to online classes or personal training sessions available through apps, YouTube videos, podcasts, and videoconferencing can be straightforward when in-person gym visits are not possible (von Humboldt et al., 2020). Many organizations offer live and pre-recorded workouts for older adults (e.g., YMCAs, Active Aging in Manitoba, 2020; Alzheimer Society of Toronto, 2021; Canadian Centre for Activity and Aging - Western University, n.d.; Osteoporosis Canada, n.d.). Bold (https://www.agebold.com/) is an online at-home fitness program specifically designed to improve older adults' balance, strength, and flexibility. After answering questions about a person's fitness level and strength, mobility, and balance, subscribers receive weekly personalized workout videos. This could be an ideal solution for Betty. With her daughter's help to teach her how to watch videos on her tablet, this strategy can facilitate her independence in an activity that is important for maintaining a healthy lifestyle.

Another technological innovation to encourage physical activity among older and younger adults alike is exergames. **Exergames** (also known as virtual reality training and exertion games) combine video game technology with sensors worn on the body or placed near the user to complete on-screen activities ranging from simulated sports (e.g., bowling, skiing) to adventure tasks (e.g., battling dragons) (Donath et al., 2016; Kappen et al., 2019; Zheng et al., 2019). Games require users to move their bodies and they provide feedback to the user so that movement and posture corrections can be made, providing both physical and cognitive stimulation (Pacheco et al.,

2020). Some are all-in-one (gaming system with controllers, sensors, game cartridge plus a screen) while others require add-ons such as a balance board or augmented reality headsets or goggles. Some commercially available exergames have a broad target market such as the Nintendo Wii Fit (https://www.nintendo.com/games/detail/wii-fit-u-bundle-wii-u/) and Nintendo Switch RingFit Adventure (https://ringfitadventure.nintendo.com/).

Exergames can be an enjoyable, engaging, and motivating way for older adults to exercise (Pacheco et al., 2020; Proffitt et al., 2015) and thus can encourage older adults to exercise regularly (Wiemeyer and Kliem, 2012). Regular use of some exergames can improve balance, mobility, and walking among older adults including those who are frail (Corregidor-Sánchez et al., 2020; Gao et al., 2020; Laufer et al., 2014; Pacheco et al., 2020; Taylor et al., 2018; Zheng et al., 2019). Older adults who have limited physical activity and do not do regular balance exercises may benefit the most from exergames. Playing such games may result in greater confidence to be more active (Laufer et al., 2014). When combined with resistance training, exergames may increase muscle strength (Zheng et al., 2019). This is significant given that older adults lose muscle strength with advancing age and are at risk for sarcopenia which can increase risk of falls (Cruz-Jentoft et al., 2019; Haynes et al., 2020). The long-term effects of exergaming on older adults' physical activity, balance, mobility, health, and quality of life have yet to be thoroughly investigated. In addition, evidence suggests that those who are new to virtual reality may experience cyber sickness (motion sickness) (Davis et al., 2014; Keshavarz et al., 2018), however other research suggests that dizziness and nausea among older adults who use virtual reality is not a major concern (Appel et al., 2020).

High tech solutions are also available to support exercise for older adults, including for those who have had an injury and require the supervision of an exercise or health professional. Similar to the idea of the Hospital at Home approach (Gonçalves-Bradley et al., 2017), these technologies bring rehabilitation into the homes of the people who require these services, including older adults. Jintronix® (https://jintronix.com/clinical-overview/) is an interactive virtual wellness and rehabilitation program that is intended for older (and younger) adults that can be implemented at home in a community setting, in assisted living, and long-term care following an injury. Using Microsoft Kinect® (a sensor that detects movement) and a screen (i.e., a television), older adults are given exercises and games that can be customized based on their specific needs by adjusting speed, range of motion, number of repetitions, and exercise duration (Lauzé et al., 2018). Information about an older adult's movements is stored and sent to the user's exercise or health professional who can adjust the routine as needed or provide feedback. The system itself can also provide the older adult with feedback in the moment to ensure that the exercise is completed correctly. A study comparing Jintronix® at home with a supervised community-based exercise program found that the home-based program can be a viable alternative to in-person programs post injury (Martel et al., 2018).

Similar systems that encourage exercise exist such as SilverFit Compact (https://silverfit.com/en; Figure 9.3), a small and portable all-in-one system comprising a computer with a built-in camera. Virtual Gym (Figure 9.4), an exergame supported by AGE-WELL, was developed using participatory design processes (Fernandez-Cervantes et al., 2018). It offers a suite of games including stretching and balance exercises led by a coach avatar and games in virtual reality in which older adults catch balloons, touch bubbles, climb a mountain, and use lightsabers. It can be personalized to complement the abilities of specific users and provides both a physical and cognitive workout for older adults.

Figure 9.3: Silverfit Compact. Used with permission.

Figure 9.4: Victor Fernandez (right), CEO of Virtual Gym, demonstrates his startup's fitness platform with Stuart Embleton. Courtesy of AGE-WELL.

Responsive Engagement of the Elderly Promoting Activity and Customized Healthcare (REACH) (https://cordis.europa.eu/project/id/690425) is an example of a high tech product designed specifically for the older adult population to support their health through physical and cognitive activity (Figure 9.5). Comprised of a consortium of 17 partners from academic institutions and industry, the goal is to transform clinical and care settings into "sensing, prevention, and intervention systems" (Hu et al., 2020). In essence, REACH supports exercise but also addresses other activities that are important aspects of improving or maintaining one's health. The system is composed of a series of smart furniture (called Personalised Interior Intelligent Units [PI²Us]) and a simulated smart home solution (TRACK) to track the activities performed by the end users. REACH is the result of a European collaboration funded by a research project under the H2020 program (Active aging and self-management of health). The REACH system has been successfully tested at the laboratory level (Hu et al., 2020).

Figure 9.5: REACH. © REACH (T. Linner, J. Güttler, R. Hu, T. Bock in collaboration with Alreh Medical, project funded by the EC's Horizon 2020 research and innovation program under *grant agreement No 690425*). Used with permission.

FIND OUT MORE

1. World Hospital at Home Community: https://whahc-community.kenes.com/mod/page/view.php?id=718

2. Virtual Gym: https://virtual-gym.ca/#about-us

3. REACH Responsive Engagement of the Elderly promoting Activity and Customized Healthcare: https://cordis.europa.eu/project/id/690425

CHAPTER 10

Technology to Facilitate Independence in Activities for Economic and Social Participation

WHAT IS IN THIS CHAPTER?

The daily lives of older adults are rich and include activities beyond self care, socializing, and enjoying life. Stereotypes of older adults include that they are warm, good-natured, friendly but incompetent, disempowered, and have reduced abilities to do everyday tasks and learn new skills (Cuddy et al., 2005; Löckenhoff et al., 2009; Makita et al., 2021). However, older adults participate economically and socially in their families, communities, and greater society through paid work, civic participation, and caregiving.

In this chapter, we outline key ways in which older adults participate in and contribute to their communities and society. We highlight how technologies can support older adults' activities for economic and social participation.

10.1 OLDER ADULTS AND PAID WORK

While there is some variation, people in many countries have shifted toward extended working lives (International Labour Organization, 2015; Phillipson, 2013). In 2017, Canadian workers were retiring, on average, at age 64, almost 3 years later than the average retirement age in the late 1990s (Hazel, 2018). In 2018, approximately one-third of Canadians aged 60 years and older were working or wanted to work in the previous year and of these, 80% worked as their main activity during a week, 10% worked but not as a main activity, and less than 10% did not work but wanted to work (Hazel, 2018). This highlights that a proportion of Canadian older adults contribute to their communities by continuing to work. For some, a continued work life during their older years is not a choice but a necessity (Ouellet-Léveillé and Milan, 2019). This is the situation in Edmund's case, described in Chapter 4. After raising four children, Edmund has little saved for his retirement. Therefore, he continues to work a physically demanding job running his family-owned construction company despite living with osteoporosis and COPD. In some countries, such as Germany, the retirement age may increase to 67 years or older (Nienaber, 2021). China has one of the lowest retirement ages at 50–55 for women and 60 for men (Kawate, 2021; The Economist, 2021).

In an era of several economic recessions, as there are fewer government- and employer-guaranteed income supports following retirement, and increased life expectancy, some older adults continue to work to have the financial resources needed to look after themselves and their families. Others choose to have a longer work life and they delay retirement because they see their work as meaningful, enjoyable, and socially rewarding (Beier et al., 2020; James et al., 2020). As David Langtry, the acting chief commissioner of the Canadian Human Rights Commission stated, "We're not born with date stamps saying our fitness for work expires at 65" (CARP, n.d.). Some older adults choose a gradual transition to full retirement by working part-time, stepping down from some aspects of their jobs, or becoming entrepreneurs, such as Jamie, the person profiled in the next section of this book (Halvorsen and Morrow-Howell, 2017). Older adult entrepreneurs, also known as seniorpreneurs (Maâlaoui et al., 2013) are self-employed and thus have the benefits of receiving an income but with greater flexibility and on their own terms (Halvorsen and Morrow-Howell, 2017).

10.1.1 TECHNOLOGY TO SUPPORT PARTICIPATION IN PAID WORK

The way in which people work and the nature of their work is changing, partly due to the introduction and integration of new technologies on a continuing basis (Czaja et al., 2019). This means that workers are faced with pressure to learn how to use new technologies so that they may do the tasks needed for their jobs. They must always update their knowledge and skills, adapting to the work world on a continuous basis. Regular upgrades to operating systems and software are now necessary, as is realizing that what was recognizable and usable to a worker one day may change and be unrecognizable the next. The very nature of how organizations operate has shifted. Teamwork, and thus the need for collaborative software such as Slack and Google Hangout, requires the constant exchange of information (Czaja et al., 2019). These demands are faced not only by older adults in paid work situations but also those who take part in some types of volunteer work, especially those with leadership and administrative responsibilities.

There is also a trend toward telework and this is expected to remain after the COVID-19 pandemic is over. Telework applications, such as Google G Suite and Microsoft Teams, permit the ability to work from home (Czaja, 2017). Telework can be beneficial to older adult workers as well as people with disabilities. It decreases (or eliminates) the need to commute and thus makes it possible for people with severe mobility challenges to continue working (Czaja et al., 2019). It may also permit greater flexibility in the work day, allowing for more breaks. The flexibility associated with telework can be helpful to those older workers with caregiving responsibilities (Czaja, 2017). Telework also brings with it a permeable boundary between work and home life leaving less time for rest as well as risk of musculoskeletal injury due to poor work station design (Czaja et al., 2019). There is also limited opportunity for social interaction with co-workers, less training, and potentially less supervision (Sharit and Czaja, 2009). For workers who were already working from home before the pandemic made it a global phenomenon, telework actually brought them into the mainstream for a period of time.

Older adult workers may face negative stereotypes. Common assumptions include that older adults are not good at using technology, do not want to learn how to use it, are not adaptable, are inflexible, and unwilling to change (Czaja et al., 2019; Harris et al., 2017). However, older workers are perceived by some as having better soft skills—such as interpersonal skills, reliability, work ethic, commitment, and organizational wisdom—than younger workers (Harris et al., 2017; Mulders, 2019; Vasconcelos, 2018). Yet, over time, negative stereotypes can be internalized resulting in a decline in self-worth and withdrawal from employment, even when income from paid work is a necessity (Harris et al., 2017). An important aspect of continued working is providing older workers with technology training so that they may acquire or improve their skills. Employers may consider providing self-paced learning opportunities and complementing these with ongoing instructions and support (Czaja et al., 2019; Czaja, 2017). E-learning modules can be supplemented with hard copies of user manuals and such support is important not only when a new technology is implemented but also when system changes and upgrades take place (Czaja et al., 2019; Czaja, 2017).

AGE-WELL is supporting a research project led by Janet Fast (University of Alberta) and designed to enhance the employability of older workers and family caregivers (AGE-WELL, n.d.). The project utilizes a platform and vocational guidance tool called MatchWork. MatchWork is a "social enterprise technology platform for economic development agencies, workforce support organizations, and governments, who are seeking to better understand their workforce. The platform helps agencies and employers connect with populations at risk of being excluded from the workforce" (MatchWork, 2020). Using machine learning embedded in the MatchWork platform combined with support from the MatchWork staff, older workers and family caregivers receive employment profiles, matches are created for employers looking to hire older workers, and workforce information is provided to community agencies who need this data to support employment strategies for workers at risk of being excluded from the workforce.

10.2   OLDER ADULTS AND CIVIC PARTICIPATION

Civic participation is characterized by active participation "… in the life of a community in order to improve conditions for others or to help shape the community's future" (Adler and Goggin, 2005, p. 241). The purpose is typically altruistic such as to "help others, solve a community problem, or produce common good" or has political outcomes (Serrat et al., 2019, p. 39). Civic participation activities can include helping specific people outside of the family, voting, donating money, signing petitions, volunteering for a charitable organization, participating in political organizations, and many others (Serrat et al., 2019).

There is general agreement that civic participation can be advantageous for older adults and their communities alike (Gonzales et al., 2015). It can decrease loneliness (Carr et al., 2018) and depressive symptoms (Choi and Bohman, 2007), positively impact cognitive function (Guiney et al., 2021; Proulx et al., 2018), reinforce one's identity (Gottlieb and Gillespie, 2008), and promote physical health (Adams et al., 2011; Burr et al., 2011; Sneed and Cohen, 2013). As such, some suggest that civic participation is "the gold standard for active and successful aging" (Serrat et al., 2019, p. 38).

Over one-third of Canadian older adults engage in some kind of formal volunteer work (Vézina and Crompton, 2012). Although other age cohorts have higher rates of volunteerism, older adults devote more hours to their volunteer pursuits than younger age groups do (Vézina and Crompton, 2012). Older adults' contributions to society through formal volunteer work has been identified as "one of the United States' most valuable yet arguably unsung resources" (Carr, 2018, p. 479). Motivations for volunteering include social connection and self-growth (Same et al., 2020), as well as goodwill, solidarity, equity, social justice, and reciprocity (Rochester et al., as cited in Grotz et al., 2020). The nature of the tasks can be diverse, ranging from providing hands-on support (e.g., delivering food, giving rides, shoveling snow, tutoring) to activities focused on advocacy and social action (e.g., community building, organizing petitions, fundraising, writing letters to politicians) to leadership (e.g., board membership, mentorship). Research participation, as a research subject, an advisory committee member, or a co-researcher, may also be considered types of volunteer activity. A growing number of older adults, such as Christine Thelker (Figure 10.1) use their lived experiences to advocate for social and policy changes (Adams et al., 2011).

People also participate informally in their communities by neighboring. Whereas volunteer work is typically viewed as a formal arrangement between a person and an organization with set expectations and procedures, neighboring is an informal arrangement associated with a particular geographic area to improve it for others and for themselves. Examples include checking in with others to ensure their well-being, keeping an eye on neighbors' properties, knowing the routines of neighbors for safety purposes, keeping an area of a neighborhood clean and sidewalks free of snow, providing support to neighbors by running errands on their behalf or driving them to appointments, and sharing information about current neighborhood events. Because older adults spend more time closer to home than in their younger years (Gilroy, 2008; Hovbrandt et al., 2007), neighboring becomes more important as people age.

Figure 10.1: Advocate Christine Thelker speaking at a conference. Used with permission.

As described, older adults may contribute to their communities and to society in many ways. Roger, who was introduced earlier in this book, participates by formally volunteering with many organizations, taking part in research as a co-researcher and advisor, and assisting those in social networks in many ways including helping them access and use technologies. However, not all older adults, particularly those who do not have the ability or interest, are active volunteers. In addition, not all older adults possess the health or social resources nor the time to be volunteers outside of their homes. Some have obligations (e.g., caregiving responsibilities) that pose restrictions on their time or may not be physically or cognitively well enough to be actively engaged as volunteers. A culture of expected productivity, in the form of volunteering among older adults, can result in the devaluing of those who cannot or choose not to participate in this way (Martinson and Minkler, 2006).

10.2.1 TECHNOLOGY TO SUPPORT INDEPENDENCE IN CIVIC PARTICIPATION

Not only has the nature of paid work changed because of the Fourth Industrial Revolution but so has the way in which volunteer work is done. In recent years, an increasing amount of volunteer work, called virtual, online, or digital volunteering, is now done virtually using the Internet rather than in person (Mukherjee, 2011). This transition to virtual volunteer work has become even more pronounced since the COVID-19 pandemic. Everyday digital technologies play a role in support-ing volunteer work. Some volunteer jobs can be carried out using these technologies. Fundraising, political action campaigns, tutoring, friendly visiting, and communication and leadership work (e.g., board and advisory committee membership) can be done using videoconferencing, social media, and email. For example, the American Red Cross has launched a volunteer campaign in which digital advocates share information about their initiatives and mobilize donation efforts (American Red Cross, 2021).

Given the health benefits of volunteer work for older adults, the disruption of in-person volunteer activities among this population has yet to be realized (Grotz et al., 2020). The possibility of shifting to virtual volunteering may be a reasonable temporary measure until such time as it is safe for in-person volunteer work to resume. Some research suggests that virtual volunteer work may be a viable option for some older adults (Lachance, 2020). Whereas formal volunteering appears to decrease with increased internet use among younger adults (i.e., internet use is seen as an alternative to volunteer work), older adults are more likely to volunteer if they use the Internet (Filsinger and Freitag, 2019). The Internet may help older adults who are keen to volunteer but have limited mobility to overcome physical barriers to volunteer work. However, virtual volunteering relies on technical skills, the financial means to purchase devices, and internet infrastructure (i.e., sufficient broadband speed). This may further exacerbate the digital divide between those older adults who have the means to benefit from virtual volunteer work and those who do not. As described in the preceding chapter and in the section pertaining to paid work, user-centred training and support for technology use is essential to support older adults to successfully participate in virtual volunteering. This *just right* training should be flexible so that it meets the needs of older adults with diverse technical abilities. Training must suit the unique needs and abilities of each older adult. Financial support from governments and policies that support access to affordable technology and the Internet is also needed to reduce barriers to participation. Social ties and social support can be effective strategies to facilitate learning about technology such as the Internet (Choi and DiNitto, 2013a).

Social benefits can also be accrued as a result of volunteering. Interacting with people and organizations can result in obtaining information about resources and services. These interactions can also result in relationships with others which may lead to providing practical support to one another. As people get to know one another, they may be willing to drive one another to their volunteer jobs or appointments or run errands for each other. Thus, information and practical support gleaned through relationships with people and organizations that developed by virtue of civic participation may help older adults maintain or improve their independence. In this way, older adults may also be a part of a larger social network of interdependence. Thus, when face-to-face interactions with others become limited due to physical limitations or public health restrictions, not only are the health benefits of civic participation jeopardized but so are the social benefits. Again, everyday digital technologies can facilitate ongoing civic participation. Moreover, support to teach older adults who have fewer social ties to use everyday digital technologies can expand not only their skills but also increase their social interactions and ties with others, potentially resulting in greater access to information and mobilizing practical support from those in their networks (Choi and DiNitto, 2013a). As a result, both independence and interdependence can be affected.

Aside from everyday digital technologies to facilitate independence in civic participation, other examples include digital organizers to maintain older adults' schedules. As some older adults report having busier schedules upon retiring than they did during their working lives, calendar systems can support them. Passive reminders, such as paper diaries, post-it notes, and calendars, are sufficient for some older adults who do not have memory challenges. However, technology that incorporates active reminders, such as an alarm or a visual alert, that prompt users to check their calendars may be helpful for those living with memory impairments (Boman et al., 2010; de Joode et al., 2012). Examples include mobile phone apps and web-based calendars (e.g., Google calendar). Roger, the person living with dementia introduced in the previous section of this book who is actively involved in many advocacy initiatives, uses Google Home, a smart assistant, to manage his schedule.

RemindMe, a digital calendar that incorporates reminders via short message service (SMS) messages sent to an older adult's cell phone, was created to overcome the drawbacks of passive reminders and existing active reminder systems (Baric et al., 2019). It was developed, tested, and improved using feedback received from older adults who can either set up the calendar and reminder system themselves or allow others (e.g., family, friends, service providers) to set it up for them. The interface was designed to be straightforward and user-friendly by limiting unnecessary options, using a simple color palette, and keeping the number of active windows to a minimum. Upon receiving the SMS message, the user responds by answering "yes" or "no" to confirm that the activity has been completed. This provides users with a record of completed activities allowing users to self-monitor their activities. Some evidence suggests that a technology-mediated active reminder like RemindMe can help older adults perform their activities, and enhance feelings of independence and control over their daily activities (Baric et al., 2019).

PRISM is a simplified computer system that can provide older adults with access to information (through the Internet and access to online classrooms), social connections (email, virtual communities, sharing photos), activities (games), and support daily activities (calendars) (Mitzner et al., 2019). Short for Personal Reminder Information and Social Management, PRISM uses off-the-shelf hardware (a monitor, keyboard, desktop PC with Windows operating system, mouse) and specially designed software to meet older adults' capabilities and needs. The interface features large font size with a simple design. Training is tailored to the needs of older adults (i.e., one-to-one, supplemented with print materials, access to a helpline). Although PRISM was developed as part of a research study for socially isolated older adults, it may be a solution for older adults who want to volunteer virtually but are novice computer and internet users.

Social platforms have also been developed to support older adults who wish to help others in their communities. A web-based community platform was co-designed by older adults in Barcelona, Spain and a group of researchers (Righi et al., 2015). This member-only platform was created to help older adults exchange support with others in their geographic region. Older adults could request support by posting the type of help needed as well as respond to someone else's post asking for help. Types of support provided and received were primarily socially oriented. Examples were diverse and included having someone to talk to, receiving technology-related help, and having a walking partner. Social platforms can enable older adults with physical limitations to participate actively in their communities. Developing relationships with others can also result in receiving the support important for maintaining well-being and living independently. This platform appears to be similar to other member-only platforms aimed at social networking such as Facebook groups for older adults in a specific geographic region or those based on special interests (e.g., food security, hobbies, sports, cultural groups) and health conditions.

10.3   CAREGIVING

In Canada and many countries worldwide, older adults can make critical contributions to the well-being of family and friends, and society in general, by caregiving. According to the 2018 Canadian General Social Survey on Caregiving and Care Receiving, approximately 25% of older adults (age 65 years and older) cared for or helped family or friends who had a long-term condition, disability, or problems related to aging (Arriagada, 2020). Older adults are most likely to spend the longest hours providing care compared to other age cohorts (Sinha, 2013). Indeed, 18% of older adult caregivers spend 30 hours per week looking after care recipients; this is nearly equivalent to a full-time job (Arriagada, 2020). With advancing age, older caregivers may provide care while also managing their own health concerns. The demands of caregiving can affect older adults differently than younger adults. In addition, young older adults (age 65–74 years) may have caregiving responsibilities for similarly aged partners and friends *as well as* their own parents *as well as* their grandchildren. For some older adults, the COVID-19 pandemic has resulted in more childcare responsibilities given the closure of daycares and schools (Meyer, 2020). Grandparenting, when older adults provide regular childcare to their grandchildren and thereby support their adult children, is a key role for some older adults (AGE Platform Europe, 2021), including Gladys, the person featured in Chapter 2.

AGE-WELL has supported a technology that can be used by older adults who are remote caregivers. Called FamliNet (Figure 10.2), this app is designed to prevent social isolation by keeping older adults in contact with family and friends. It is a multimedia messaging system that can bring together a circle of support (family and friends) that surrounds an older adult to enable group, as well as one-on-one, communication. FamliNet is compatible with all major devices and has been designed to accommodate older adults. It features large text size and easy navigation, converts voice messages to text messages, and vice versa. This allows older adults who have difficulties with dexterity or vision to speak a message into the app which is then converted to text. For those with hearing impairments, voice messages can be converted to text messages, expanding opportunities for communication among older adults who have hearing difficulties. FamliNet also includes an easy-to-use videoconference function. It facilitates older adults, who may or may not be technologically literate, to be remote caregivers to family and friends who may likewise have limited technology skills. For Betty, the person featured at the beginning of this section of the book who is a novice technology user, FamliNet could allow her to spend time online with her grandchildren during the COVID-19 lockdown. While not replacing face-to-face interaction, FamliNet may help her maintain her role as grandmother by reading, visiting, or playing with her grandchildren online. Although being there to provide hands-on help to her daughters by looking after her grandchildren would be the ideal, spending time with them online could still be helpful and enjoyable.

Figure 10.2: Steph Gagne and her father, Richard Ratcliffe with FamliNet, an app that offers older adults an easy way to connect with family and friends. Photo by John Hryniuk. Courtesy of AGE-WELL.

FIND OUT MORE

1. Famlinet: https://www.famlinet.com/

CHAPTER 11

Technology to Facilitate Enjoyment and Self-Fulfillment

WHAT IS IN THIS CHAPTER?

In this chapter, we describe how technologies can support older adults' participation in activities for the purposes of enjoyment and self-fulfillment. We introduce these types of activities and their roles in the lives of older adults and incorporate examples of technologies that facilitate participation and independence in these activities. We conclude by describing issues concerning technologies to support older adults' independence in such activities.

11.1   WHAT ARE ACTIVITIES FOR ENJOYMENT AND SELF-FULFILLMENT?

Regardless of one's age, life is about more than looking after oneself (Chapter 6) and contributing to one's community and society (Chapter 7). People also spend time doing activities for enjoyment and self-fulfillment. Known as **enhanced activities of daily living** (Rogers et al., 1998, 2020), these activities are done for the purposes of fun, pleasure, relaxation, learning or growth, play, leisure, recreation, and entertainment (Figure 11.1). Taking part in these activities can be personally enriching, add meaning to one's life, and meet higher psychological and spiritual needs than activities that focus on self care. Unlike self care and community contribution activities, older adults may have greater choice and freedom in selecting these activities; they are voluntary as opposed to obligatory. Whereas self care and community contribution activities are must-dos (e.g., ADLs, IADLs, caregiving, paid work) or should-dos (e.g., volunteering, looking out for neighbors), activities for enjoyment and self-fulfillment are *want* to dos in which the older adult's preferences are expressed, offering a break from other activity types. Although not an element of all activities for enjoyment and self-fulfillment, many involve aspects of social interactions and relationships (Rogers et al., 1998, 2020). In essence, activities that center on enjoyment and self-fulfillment are satisfying and meaningful to a person and can add life to years.

Figure 11.1: Characteristics of activities for enjoyment and self-fulfillment.

The COVID-19 pandemic has brought to the forefront activities done for the sake of enjoyment and self-fulfillment. With stay-at-home directives and restrictions on social gatherings, some have considered how they currently spend their time and have taken up new hobbies or embarked on learning opportunities. The salience of leisure in our everyday lives has been given more attention than in the past (Lashua et al., 2021) and the attention that it deserves. After all, such activities are an important part of aging well (Gow et al., 2017; Kuykendall et al., 2015; Paggi et al., 2016; Yates et al., 2016). Thus, some suggest that helping people participate in these activities should be a focus for health professionals such as occupational therapists (Chen and Chippendale, 2018). Yet, the notion of leisure is not universal (Hammell, 2009). Moreover, not all older adults have the time or resources to participate in activities focused on enjoyment and self-fulfillment. Some may have little free time due to other responsibilities such as paid work and caregiving. Others may not have the financial means nor physical and cognitive abilities to engage in such activities.

11.2   TYPES OF ENJOYMENT AND SELF-FULFILLMENT ACTIVITIES

There is no typical older adult. This demographic is heterogenous, consisting of at least three different cohorts as well as diverse abilities, cultures, ethnicities, social and economic resources, values, and preferences. As such, the activities that they participate in for fun and self-fulfillment are as diverse as older adults are. What may be meaningful and enjoyable for one older adult may not be for another. However, some general statements can be made about how older adults spend their time. The nature of their activities for enjoyment and self-fulfillment are similar to those of younger adults and follow patterns established earlier in life (Czaja et al., 2019). For example, watching television is the most common leisure activity among all age groups (Czaja et al., 2019). Older adults spend more time taking part in passive leisure activities each day (e.g., reading, watching TV, talking on the phone, listening to the radio) than active leisure activities (e.g., exercising, socializing in a club, using technology, traveling, volunteering) (Arriagada, 2018; Cho et al., 2018). Yet, approximately 75% of Canadian older adults aged 65 and older report spending approximately 3.5 hours per day doing active leisure activities (Arriagada, 2018). Not surprisingly, older adults who are not employed have higher participation rates than those who are employed (Arriagada, 2018) indicating that how older adults spend their time changes upon retirement and indeed, attention must be given to these activities focused on enjoyment and self-fulfillment rather than only those associated with self care. Leisure activities may also create a sense of purpose and meaning following a transition in older adults' roles and routines after retirement from paid work (Nimrod et al., 2012).

Despite the diversity of older adults and their activities, for the purposes of this chapter, we have grouped their enjoyment and self-fulfillment activities into the seven categories depicted in Figure 11.2. Technologies to support independence in these activities are described in the section that follows.

There are undoubtedly many other types of activities that older adults participate in, and they vary by culture, ethnicity, country, social and economic resources as well as individual preferences. Most notably, watching television and movies and listening to music are very popular activities. Participation in these activities can be supported using smart home technologies in which users control their entertainment media using voice controls, a tablet, or voice controls. Similarly, digital voice assistants (e.g., Google Home, Alexa) can be set up so that those with physical or visual challenges can listen to their favorite music without having to negotiate their home environments. Another technology to support listening to music is the Simple Music Player, designed to help persons living with dementia to select and listen to music of their choice (Sixsmith et al., 2010; Sixsmith and Gibson, 2007). Notably, this technology was part of a large research project, entitled INDEPENDENT, that involved co-designing, developing, deploying, and testing devices to enhance the quality of life of people living with dementia (Orpwood et al., 2007; Sixsmith et al., 2007).

Figure 11.2: Types of activities for enjoyment and self-fulfillment.

11.3   TECHNOLOGIES TO SUPPORT INDEPENDENCE IN ACTIVITIES FOR ENJOYMENT AND SELF-FULFILLMENT

As described in Chapter 8, an abundance of technologies has been developed to support older adults' independence in self care activities. There is less research on technologies to support older adults' participation in activities for enjoyment and self-fulfillment (Boot, 2020). This is of concern given that these activities are important for aging well. Moreover, when technologies are developed for older adults, they may be evaluated with healthy older adults not those who have health symptoms (Poli et al., 2020). This could lead to results that are biased.

11.3.1   LIFELONG LEARNING AND EDUCATION

For some older adults (but not all), retirement marks the beginning of a period of personal growth replete with learning activities. These activities can be formal, such as enrolling in a for-credit or not-for-credit course or "third age" programs offered by post-secondary institutions. First established in the early 1970s, the Universities of the Third Age movement refers to organizations, often affiliated with universities, that provide education specifically for older adults (Formosa, 2012; Ratsoy, 2016). The University of the Third Age (U3A) is an international organization with member groups in 23 countries including a Virtual U3A division (Ratsoy, 2016). The purpose of U3As is to provide intellectual stimulation and social interaction to increase quality of life among participants. Subject matter can range from history to art to religion to health to creative writing. In Canada, universities affiliated with U3As include Ryerson University's Chang School of Continuing Education (https://continuing.ryerson.ca/) and Simon Fraser University's Liberal Arts and 55+ Program (https://www.sfu.ca/continuing-studies/programs-and-courses/area-of-study/liberal-arts/courses-and-events/online.html). In the United States, the Stanford Careers Institute (https://dci.stanford.edu/) "fosters intergenerational engagement in an academic setting to help create a new paradigm for the university of the future." Intergenerational interactions can help younger adults develop confidence that their lives matter to others, recognize vulnerabilities, and learn from the life experiences of older adults (Carr and Gunderson, 2016). For some older adults, such programs give an opportunity to take part in courses or complete degrees (including graduate degrees) that they perhaps had an interest in but never had time to participate in earlier in their lives. For others, it is time for them to create new relationships that last the rest of their lives.

There are also community-based, less formal opportunities for lifelong learning that leverage technology and aim to stimulate and strengthen social connections. Seniors' centres and organizations offer online classes in which older adults can learn a new hobby, expand their knowledge, or receive practical information on topics ranging from wellness to income taxes. Courses can be synchronous (e.g., Zoom videoconferencing) or asynchronous (e.g., instructional videos). Topics range from how to paint (hosted by Creative Aging Calgary Society, 2021) to how to sing in a choir and these retain an element of bringing together older adults for the purposes of learning and socialization. Low tech options are also available. The Temiskaming Seniors Centre without Walls (https://www.timiskaminghu.com/80471/seniors-centre-without-walls) provides a free, telephone-based program including learning opportunities and a support for older adults age 55 and up, particularly for those who are socially isolated. SPF Seniorerna (the Swedish Association for Senior Citizens; https://www.spfseniorerna.se/about-us-in-english/) and PRO (https://pro.se/om-oss.html) organize learning and social opportunities for older adults.

Other lifelong learning and education opportunities with less of a focus on building a community of older adult learners include Massive Online Open Courses (known as MOOCs). Many universities, including Yale and Stanford, offer MOOCS. These range from free of charge to low cost, delivered on online platforms (Petrecca, 2019). The wide array of subjects includes artificial intelligence, paleontology, Indigenous history, and astronomy, providing older adults with a potpourri of content to select from. MOOCs offer educational opportunities to older (but also younger) adults that they may not have been able to access in person due to financial resources, geographic location, and physical abilities (Schmid et al., 2015).

Many of the opportunities described rely on everyday digital technologies. And, a digital divide remains for older adults and marginalized segments of our population. Not all older adults have access to everyday digital technologies and a reliable internet connection. If they do, they may have insufficient skills or confidence to be able to participate in virtual lifelong learning and education opportunities. Still others may require intermittent support to set up and use these technologies as well as troubleshoot when difficulties arise. Age-related hearing and vision changes may impact older adults' ability to successfully take part in virtual learning opportunities, as may changes in motor and cognitive skills needed to use everyday digital technologies. Furthermore, the cost of enrolling in some programs, such as U3As, may be a barrier to participation. Thus, while there are many lifelong learning and education opportunities available to older adults that can be facilitated using everyday digital technologies, it is likely that access favors older adults who are affluent and not marginalized.

11.3.2 SPORT AND PHYSICAL ACTIVITIES

Contrary to popular assumptions that paint an image of older adults as primarily sedentary, involvement in physical activities can increase after they retire (Czaja et al., 2019). In fact, older adults aged 65 and older identify four of their top five activities as being physically active and these include walking, outdoor maintenance, playing sports, and other physical activities (Szanton et al., 2015). Even adults over the age of 80 describe three of their top five activities as being active (Szanton et al., 2015). However, some older adults may interpret exercise and physical activities as "work" (i.e., "must-dos" to maintain their health) while others may view these activities as being enjoyable and fulfilling depending on the nature of these activities. For example, skiing, golf, and tennis may be seen as fun whereas going to the gym may be regarded as less enjoyable and less meaningful.

Sport and physical activities may be impacted by changes in physical abilities secondary to chronic conditions that may increase with age. This means that some older adults may find it increasingly difficult to maintain participation in sport as they get older. Also, stay-at-home orders, the stop in organized sport activities for older adults, and the closure of recreation facilities have affected older adults' opportunities for physical activity. Whereas prior to COVID-19 older adults may have been limited by their physical and cognitive abilities and resources, older adults now experience more barriers to participation and alternatives to traditional forms of sport are needed (Son et al., 2021). Technologies and technology-based initiatives that do not require in-person contact may be a potential solution to address these barriers to participation.

An example of an innovation that can support older adults' independence in sport are pedelecs (known by some as electric bikes or e-bikes; Figure 11.3). Pedelecs are equipped with an electric motor that, when activated, can assist older adults with pedaling. This reduces the physical demand on older adults, especially when navigating hills, and can help those who have reduced abilities to participate in cycling. Pedelecs may enable older adults to cycle for longer distances and with less effort which can make the activity more enjoyable (Van Cauwenberg et al., 2019). Some evidence also suggests that exercising in an outdoor environment (as opposed to indoors in a gym) may have added benefit for older adults. Cycling using an pedelec may help older adults not only to accrue the health benefits of physical activity but doing so outdoors (Leyland et al., 2019). Even though less physical effort is required by older adults in comparison to a conventional pedal bike, using a pedelec can still have a positive impact on well-being and cognitive function (Leyland, 2016). The benefit can be even greater when outdoor activity is coupled with social interaction, and evidence suggests that older women in particular use pedelecs for social reasons (Van Cauwenberg et al., 2019). Pedelecs are more commonly used by older adults in Europe than in North America; barriers to adoption including cycling infrastructure and road safety, regulations, perceived stigma around older adults using pedelecs, fear of falling or injury, bike weight, and concerns about battery life (Leger et al., 2019; Van Cauwenberg et al., 2019).

Figure 11.3: Cycling as an enjoyable activity.

Older adults who have greater physical and cognitive limitations and thus live in supportive living settings like a continuing care centre or retirement home can also be supported to participate in cycling. BikeAround consists of a pedaling unit, handlebars, a software system, and either a television screen or a specialized dome containing a projector (Camanio Care, n.d.). Using Google Street View, BikeAround allows older adults to virtually cycle through familiar places as well as others that they may have always wanted to visit. It provides exercise while also facilitating cognitive stimulation and reminiscence (Larkin, 2018). The system containing the dome projector can be deployed in assisted living or continuing care settings whereas the simple version containing a television screen can be used in older adults' private homes in community settings. Research is needed to evaluate the impact of BikeAround on older adults' mobility, quality of life, and perceived health; to date, it has been the focus of only a small-scale study that explored the perspectives of nursing staff (Shen, 2020). A variation of BikeAround is Motitech which is currently being trialed in some long-term care facilities in Ontario, Canada (Motitech, n.d.).

An emerging area of innovation and research is robotic apparel, also known as robotic exoskeletons. Such wearable robotics are flexible suits that are worn on the trunks, arms, and legs. These suits contain sensors and electric motors that can provide strength and stability to movements. Users still have to initiate movements, but the suits provide assistance so that movements are easier and take less energy. Robotic exoskeletons are currently used for rehabilitation purposes, are heavy and bulky, and require trained personnel to help older adults use these devices (Panizzolo et al., 2019). Examples include Rex Bionics (https://www.rexbionics.com/rex-for-clinical-use/), ReWalk (https://rewalk.com/) and Cyberdyne HAL (https://www.cyberdyne.jp/english/products/HAL/). The AXO-SUIT (https://www.axo-suit.eu/) is an exoskeleton developed through collaboration between several European universities and companies. It is intended to support older adults in their daily activities including leisure. A lightweight wearable robot that has been integrated into a textile has been developed to assist with walking. This prototype has been tested with older adults in a proof of concept study (Panizzolo et al., 2019). A soft, wearable ankle robot prototype was also recently developed and tested with older adults with success (Koo et al., 2020). These developments show promise and suggest that in the future, lightweight and commercially available wearable robotics may help older adults to participate in the sports and physical activities that they enjoy and value.

As previously described, everyday digital technologies such as computers, tablets, cell phones, combined with internet access or applications can also help older adults take part in sport activities for the purposes of fun and self-fulfillment. With the help of her daughter, Betty (the older adult profiled at the beginning of this section) is able to access Tai Chi videos on YouTube. Studying Tai Chi has been a goal of Betty's for some time but given her difficulties with transportation, it was difficult for her to access classes in her community. Although she finds it not the same as taking part in a physical activity in person, using Tai Chi videos allows her to keep active during the current COVID-19 lockdowns. Other physical activities are available online, ranging from dance, to yoga, to Zumba, to Pilates.

For Roger (the person living with dementia featured at the beginning of the section on autonomy) who has a lot of experience with everyday digital technologies, using the Internet can help him access information about his community such as facilities offering specific programming, groups that meet up to go mountain biking, or the best golf courses to visit in a region. This, in turn, can help support his participation in these sport-related activities. Given that the Internet can provide geographically specific information to older adults which can facilitate their activities, there is a need for programs that target older adults' digital skills and that create platforms for information sharing (Schehl, 2020).

11.3.3 READING

Some older adults, particularly those who are not employed, report spending their free time reading for pleasure and their participation rates are higher than younger adults (Arriagada, 2018; Czaja et al., 2019). Although people in all age groups prefer print books over e-books, there is still a divide between older and younger adults as it relates to accessing books electronically (Czaja et al., 2019). Reasons may include lack of access to devices (i.e., e-readers, tablets, smart phones) and an internet connection, lack of education, training, and confidence in how to use devices and e-book apps, and visual challenges that may interfere with the ability to discern text or navigate menus. Yet, there are advantages to older adults using e-books. For example, font size and contrast can be adjusted on some devices, and most e-books provide an audio option. Some libraries provide access to tens of thousands of e-books to their members free of charge; older adults thus have access to a plethora of reading material without having to leave their homes. This improves access for those older adults with mobility and transportation limitations. Everyday digital technologies can also provide access to audiobook player apps which can increase independence in reading for those older adults who have significant visual challenges.

Figure 11.4: When Rachel Thompson couldn't find "dementia-friendly" books for her grandmother, she created her own specialized books—and founded Marlena Books in 2016. The stories are all written by Canadian authors. Courtesy of AGE-WELL.

Marlena Books (https://marlenabooks.com/pages/what-makes-us-different) is a novel technology that supports persons living with dementia to participate in reading (Figure 11.4). The company offers print books as well as a digital book platform. Books in their collection are age-appropriate reading material for adults and contain content that is appealing to older adults. The Marlena Books mobile application is available for Apple IOS devices and was specifically designed to meet the needs of persons living with dementia. Accessibility features include automatic page turning, large font, and a simple navigation menu. Marlena Books can provide persons living with dementia the opportunity to maintain independence and enjoyment in reading despite their changing cognitive abilities.

11.3.4 SOCIALIZATION AND COMMUNICATION

Next to watching television, activities centered on socializing and communicating are the next most common activities for enjoyment for older and younger adults alike (Czaja et al., 2019). And, like younger adults, older adults use everyday digital technologies combined with apps, software, and social media platforms to maintain relationships with family, friends, and colleagues, especially when facing physical distancing restrictions, self-isolation, or geographic barriers (von Humboldt et al., 2020). Also similar to younger adults, some older adults use apps for online dating and to find partners later in life (Marston et al., 2020). Although not without its risks and challenges (Age UK, 2021; Field, 2018; Marston and Kowert, 2020), dating apps can help older adults looking for relationships to meet others which can become increasingly difficult when there are fewer in-person opportunities or it becomes more difficult to leave home due to health conditions. Videoconferencing, supported through Zoom, Skype, WhatsApp and other platforms, can also enable older adults to connect with others in their social networks for one-on-one or group conversations. Some senior centers arrange Zoom meetings in which older adults can visit and some of these are organized around an activity. For example, the Edmonton Seniors Centre offers a virtual get-together that centres on knitting and crocheting (www.edmontonseniorscentre.ca/product/knitters-crocheters-unite-may/).

Older adults, particularly those who are younger, female, and have higher education and income, also use social networking sites for social engagement and fun, and it is expected that use of social media among older adults will continue to increase (Anderson and Perrin, 2017; Hunsaker and Hargittai, 2018; Leist, 2013; Smith and Anderson, 2018; van Deursen and Helsper, 2015). Older women may use social media to stay informed about their children and grandchildren's lives and Facebook's ability to maintain social connections may outweigh its negative aspects, such as privacy and advertising (Hebblethwaite, 2017). Older adults may also use social media to provide social support to others and to receive social support, especially from those who have similar interests, backgrounds, and life experiences such as caregiving issues, mental health concerns, and specific health conditions (Leist, 2013). The flexibility of providing access to others synchronously as well as asynchronously allows people to connect despite differences in time zone, their availability, geographic barriers, and abilities. Importantly, use of online communication mediums supported by the Internet can expand social networks and may produce in-person interactions that would otherwise not have occurred were it not for online communication (Russell et al., 2008). For Marni, the person profiled in Chapter 1, a digital presence is a medium to facilitate sharing her story and fostering understanding of experiences of people with different gender identities.

As described in Chapter 10, some communication and socialization apps are designed specifically for older adults who have limited experience with social media platforms and the Internet. These apps provide a streamlined interface with fewer non-essential options, larger font and better contrast, and no advertisements. FamliNet (https://www.famlinet.com/) is such an example and although it was developed to support family caregiving, it can also be used to facilitate socialization and communication among older adults. It offers voice-to-text and text-to-voice transcription, supporting those with visual or hearing impairments who may find it difficult to use widely available social apps.

Social robots present another option to support older adults to participate in activities pertaining to socialization and communication. Social robots, or "socially interactive robots," are high tech solutions capable of having meaningful interactions with people (Sandry, 2015; Sarrica et al., 2020). Human-like social robots have been applied in research settings with persons living with dementia. Mario is a robot who interacts with people by reading newspapers to them, playing music, sharing photos, helping them connect remotely with family and friends, and reminding them of appointments (Mario, 2015). It responds to voice commands and also has an onboard touch screen to control its interactions. Project ENRICHME (ENabling Robot and assisted living environment for Independent Care and Health Monitoring of the Elderly) is a humanoid, multi-purpose robot that can be used for health monitoring, giving reminders, interacting with people, and playing games (Coşar et al., 2020). Unfortunately, both Mario and ENRICHME are only available for research purposes; they are expensive, difficult to set up, and are not available on the market. However, two humanoid robots are commercially available: NAO and Pepper. These robots are designed for older adults and can help them to communicate with family and friends by utilizing onboard phone and videoconferencing (Kintsakis et al., 2015; Reppou et al., 2016). They interact with users by sharing news, weather, and recalling memories as well as encourage them to do cognitive and physical exercises. Like Mario and ENRICHME, NAO and Pepper contain various sensors that facilitate health monitoring and detect behaviors (Figure 11.5). However, these products have not been well studied in everyday situations in community and residential care contexts and they are costly.

Figure 11.5: Pepper social robot. Photo from Softbank Robotics Europe, CC BY-SA 4.0 via Wikimedia Commons.

Petbots (animal-like robots) have also been developed and used with persons living with dementia (Ozdemir et al., 2021). Paro (a robotic seal) (Paro, 2014), and the Joy for All dog and cat (Joy for All, 2018), are commercially available products that have been used in continuing care settings in Japan, Denmark, and the United States. While not attempting to emulate social interactions with human beings, petbots can support older adults with severe cognitive impairments to enjoy themselves. These petbots contain sensors that allow them to learn and react to users' movements and sounds. Petbots may take persons living with dementia back in time to when they had a pet of their own, thus encouraging reminiscence. This may offer comfort, enjoyment, and meaning to the user. Although these products are high technology, they do not require the user to do anything to operate them. Thus, petbots are technologies that support independence in enjoyable and self-fulfilling activities among older adults who may have very limited physical, cognitive, and communication abilities.

Technologies that support socialization and communication among persons living with cognitive impairment by providing memory and conversational prompts also exist. Known as a memory prosthesis (Gelonch et al., 2019), these devices can facilitate interactions with family and friends, helping to maintain connection and enjoyment in interacting with one another despite memory and communication challenges. The Narrative Clip® (http://getnarrative.com) is a small, lightweight, lifelogging camera that users wear around their neck. It can take hundreds of pictures per day. Pictures can be viewed on a computer by connecting the camera to a USB port, allowing users to review the events of the day. Doing so can help older adults recall their daily activities and these can be used as content for conversations with family, friends, and care providers. Earlier renditions of lifelogging technologies, such as the SenseCam, exist that provide persons with cognitive impairments a digital record of their activities and thus serve as memory technologies (Crete-Nishihata et al., 2012). It was found that reviewing pictures generated through a lifelogging camera could support conversation, reminiscence, and the sharing of stories, as well as contribute to maintaining identity in the face of cognitive impairment.

11.3.5   SPIRITUALITY

Activities pertaining to spirituality and religion play an important role in the lives of some older adults. Spirituality and religion can provide meaning, purpose, hope, comfort, connectedness, structure, and understanding (Malone and Dadswell, 2018; Manning, 2013; Rote et al., 2013; von Humboldt et al., 2020). Spiritual well-being and prayer contribute to older adults' well-being and quality of life (Agli et al., 2015; Lavretsky, 2010; Lucchetti et al., 2019). Technologies can support independence in activities pertaining to spirituality and religion.

Mobile applications, videos, and resources available on the Internet are available to facilitate older adults to engage in spiritual and religious activities. A search of the Google Play and the Apple App stores resulted in finding many apps from various religions (e.g., Hinduism, Islam, Sikhism, Judaism, Buddhism, Christianity, Jainism, Bahá'í, Zoroastrianism) in different languages that guides users through prayers, chants, hymns, scriptures, devotions, and teachings. Religious activities and celebrations can be live streamed, viewed, or listened to via podcasts. Some older adults identify these as important resources to maintain active involvement in their faith when they are unable to attend services in-person (von Humboldt et al., 2020). Apps are also available to support meditation and mindfulness such as the Insight Timer (2021) and the UCLA Mindful (UCLA Mindful Awareness Research Center, 2017).

Given the sheer volume of apps available for spirituality, let alone for people's many other interests and activities, it can be difficult to navigate app stores, especially for people with cognitive impairments. As such, a person-centred tool (i.e., an app to identify appropriate apps) was developed using participatory design approaches that sought feedback from persons living with dementia, their caregivers, and app designers and developers (Kerkhof et al., 2019). This tool is currently being tested in the Netherlands (Neal et al., 2021). In Canada, AGE-WELL has supported the creation of the Alberta Rating Index for Apps (ARIA), an index for rating the quality of health apps (Azad Khaneghah, 2020).

Communication technologies such as email, text messaging, and social media also support participation in spirituality and religious activities because these technologies bring people together. Spiritual and religious practices contribute to a sense of community and belonging (Malone and Dadswell, 2018). Among older adults, contacts through religious organizations is the second greatest source of social support, the first being family support (Koenig, 2012). Those without access to technologies that support connections with members in their spiritual and religious communities are at a disadvantage; not only can they not access the resources available online and relevant apps that could be of use to them, but they are also limited in their opportunities to connect with others and exchange social support. For members of the LGBTQIA2 communities, it is not uncommon to be disconnected from their biological families. In such cases, it is important to access technologies that facilitate social connection and support within communities where there is a sense of belonging. More spiritual communities are now inclusive of individuals who identify with LGBTQIA2, although many are still excluded and live with the trauma experienced decades past.

11.3.6 GAMING

With aging of the baby boomer generation, technology use by older adults is changing. Despite the popularity of gaming, substantially fewer older adults play video games than younger people (Boot, 2020). In recent years, gaming (i.e., playing video games) has become of interest in research as well as in everyday life. Video games created for older adults are often designed for "serious" purposes. Older adults are encouraged to play these games to facilitate their rehabilitation, for brain fitness, and for physical fitness (see Chapter 9 for a description of exergames) (Boot et al., 2020). Serious games are marketed as being good for older adults and a means to improve or maintain their functional abilities and look after themselves. However, less attention has been given to video games for older adults simply for the purpose of enjoyment.

Video games have become more easily accessible by non-gamers and thus are becoming more popular among women and adults (Czaja et al., 2019). A gaming console (e.g., Nintendo, PlayStation) is no longer the only medium for video games; they can be played on tablets and smart phones. This expansion in popularity to other groups has been augmented by the presence of *casual video games*, games that are easy to learn and play and with few rules (Czaja et al., 2019). Puzzle video games (e.g., Tile-matching, Bejewelled) appear to be preferred among older adult populations (Blocker et al., 2014; Chesham et al., 2017). Furthermore, older adults who play video games do not necessarily need to have a lifelong interest in playing games in general; one study found that playing tabletop games (e.g., board games, puzzles, cards) was not predictive of playing video games among older adults (Blocker et al., 2014).

Video games may offer advantages over conventional tabletop games for some older adults, supporting independence in activities for enjoyment. Some video games can respond to older adults' cognitive capabilities, providing the "just right" level of challenge. This allows people with and without cognitive impairments to successfully play a game and feel a sense of accomplishment but also provide a sufficient level of challenge to retain the user's interest such that the user is immersed and highly engaged in the game. In fact, an entire body of literature exists pertaining to the development and application of video games for persons with dementia or cognitive impairment.

Box 11.1 describes machine learning and how this can be applied to video games for persons with cognitive impairments to support them to play video games.

Box 11.1: What is machine learning and its potential uses? Author: Melika Torabgar (M.Sc. student, University of Alberta).

Machine learning, as a part of artificial intelligence, is a science of getting computers to learn, predict, and make decisions, similar to humans (Alpaydin, 2020). The machine, i.e., computer, is taught with different algorithms to learn and make accurate predictions of the future or classifications of events, things, or even people! Machine learning needs the guidance of humans to teach it to make the decisions from data labelled by humans. This is called supervised learning. However, the machines can also learn by themselves, called unsupervised learning; they can analyze data and teach themselves how to make decisions based on how data have behaved in the past (Graham et al., 2019). Reinforcement learning, another type of machine learning, could help the machine learn from its surroundings by giving rewards or punishments for its actions. This method is used for navigation systems in robots by doing "good" or "bad" actions and receiving feedback from the environment (Knox and Stone, 2009).

Thanks to state-of-the-art technology, machine learning can be employed in health care settings. Using data from patients, machine learning can help health care practitioners detect and treat disease more efficiently, improving patient outcomes. For instance, machine learning could help recognize cognitive decline in older adults based on the analysis of their speech, their movements and interaction with objects at home, or their performance while playing a mobile game.

Another potential use of machine learning is in recognizing engagement in older adults with mild cognitive impairment and dementia. Engagement in cognitive activities such as mobile games could reduce the speed of cognitive decline. Older adults' engagement is usually assessed with questionnaires; however, older adults who have difficulty thinking, focusing, or remembering may find questionnaires challenging to use (Wickins et al., 2019). Machine learning can measure engagement during gameplay, but data used for teaching the machine has come from younger people (between 8 and 55 years of age). Compared with young people, older adults can be different in their facial expressions, vocalizations, and verbalizations when they are engaged. Thus, there is a need to explore automated ways such as machine learning to recognize engagement in older adults with and without dementia. This requires data from older adults to teach the machine (Schrader et al., 2020). If machines can help us recognize the level of engagement of an older adult, it becomes possible to adjust the game challenge level to enhance the level of enjoyment and benefits of game activities among older adults.

Video games also give rise to social interactions either in-person or virtually. For older adults, the benefits of playing games include connecting with family and mitigating loneliness (Kaufman et al., 2019). The ability to interact with other people online, particularly family and friends, is of particular relevance in the height of the COVID-19 pandemic. Research suggests that playing video games can promote social connections among older adults, both with those individuals with whom they have local and distant ties (Lee, 2019). In essence, video games can offer older adults an opportunity to connect with people through play which is particularly important during times of stress (Marston and Kowert, 2020).

11.3.7 CREATIVE ENDEAVORS

Humans are creative beings; we participate in a range of creative activities such as making music, creating art, moving our bodies, and performing, to express our ideas, learn, have fun, and connect with others. The importance of creative expression in aging has been given prominence in the creative aging movement (Klimczuk, 2016). This movement encourages creative aging through health and wellness programs (e.g., integrating art therapies into care settings), encouraging older adults to participate in creativity-focused civic activities such as volunteer work and mentoring, and lifelong learning opportunities centered around the arts (Klimczuk, 2016). Evidence suggests that some older adults can benefit from participating in creative endeavors as they age. For example, choir activity is associated with social engagement, positive mood, and quality of life (Pentikäinen et al., 2021).

As with other activities for enjoyment and self-fulfillment presented in this chapter, everyday digital technologies can support older adults' participation in creative endeavors. Classes can be delivered through live videoconferencing, YouTube videos, and mobile applications. Baycrest and the National Ballet School in Canada have collaborated to deliver dance classes for older adults using a specially designed app (Canada's National Ballet School, 2021). Classes are taught by professional dance teachers and are available to people who prefer seated as well as standing dance positions, making the program accessible to older adults of varying abilities. Apps are also available to support creative endeavors for persons living with dementia. SingFit is a karaoke-style app available for iOS devices that can be implemented by caregivers of persons living with dementia (Musical Health Technologies, 2021). It does not require reading, making the app ideal for people who have challenges with vision or the ability to read. A professional version of SingFit is available for use by professional caregivers and therapists in group settings. Technologies are also available to support participation in and viewing of visual arts for persons living with dementia such as apps (e.g., Mattson, 2015; Tyack et al., 2017) as well as high tech devices (e.g., Leuty et al., 2013). Regarding the latter, Engaging Platform for Art Development (also known as ePAD) is a touch-screen device that uses AI to support people living with dementia to create art.

11.3.8   ISSUES PERTAINING TO TECHNOLOGIES TO SUPPORT ACTIVITIES FOR ENJOYMENT AND SELF-FULFILLMENT

In the preceding sections, we described some of the major activities that some older adults participate in. Participation in some activities for enjoyment and self-fulfillment, such as travel, looking after pets, gardening, and being in nature, cannot be facilitated using technologies; these activities rely on "in-person" interaction, take place in environments that are not conducive to the application of technologies, or cannot be replicated virtually. However, some older adults may still use everyday digital technologies combined with the Internet to inquire about, plan, write about, and buy equipment or supplies related to activities for enjoyment and self-fulfillment including travel (Nimrod, 2009; Patterson et al., 2011). It is important to recognize that access to online shopping, infrastructure, systems, and processes varies from country to country (and perhaps even from region to region). For example, in Sweden, BankID (https://www.bankid.com/en) is commonly used for online shopping. This system requires a user to have a newer model smart phone. However, not everyone trusts the BankID system and not all older adults have access to smart phones, thus limiting their ability to use online shopping to support their activities for enjoyment and self-fulfillment.

By participating in activities pertaining to enjoyment and self-fulfillment, people are exposed to others. Doing so, whether in person or online, can increase their social networks and may result in exchanging social support (Choi and DiNitto, 2013b). Also, social relationships can inspire participation in leisure activities and greater participation in leisure can result in better health and the ability to better overcome difficult life events (e.g., loss of a loved one) (Chang et al., 2014; Janke et al., 2008). Thus, when older adults do not have access to everyday digital technologies or lack the training, confidence, and skills to use these technologies, they are disadvantaged; they are subject to the "grey digital divide" (Son et al., 2021). Gladys, who is featured in a persona in Chapter 2, has limited use of everyday technologies, due to this divide. Not only is participation in activities for enjoyment and self-fulfillment limited, but so are their social resources. These social resources are potentially an important source of support to maintain independence and interdependence with advancing age. Older adults who are particularly subject to the grey digital divide are those in the oldest old age group (85 years or older), persons with low income, low educational attainment, and who live in geographic areas with poor internet connectivity (Hodge et al., 2017; Schumacher and Kent, 2020). It is therefore critical to consider programs, services, and policies to support the acquisition of everyday technologies and internet service for older adults as well as provide the skills necessary for their use. Novel, intersectoral partnerships between organizations that serve older adults could be developed to deliver such support (Son et al., 2021). Partnerships could also be developed to offer appropriate programming for older adults, particularly in municipalities with fewer resources, such that older adults who are online can access a range of programming regardless of where they live or their transportation limitations (Son et al., 2021).

FIND OUT MORE

1. Simple Music Player: http://www.dementiamusic.co.uk/

2. The University of the Third Age: https://sources.u3a.org.uk/

CHAPTER 12

Technology for Independence in Mobility in the Community

WHAT IS IN THIS CHAPTER?

Community mobility, defined as "…moving self in the community and using public or private transportation" (American Occupational Therapy Association, 2002, p. 620) can be an important aspect of maintaining independence. Some older adults participate in activities in their communities including paid work, civic activities, and caregiving which were described in a previous chapter. Leaving one's home and moving around one's community may also be an element of self care activities (such as attending medical appointments, grocery shopping) and activities for enjoyment and self-fulfillment described in Chapter 11. With advancing age, people may experience health conditions and changes in their physical and cognitive abilities that may impact their ability to safely enter and move around their communities. Yet, technologies are available to support older adults' community mobility and, in turn, support their participation and independence in activities.

In this chapter, we provide descriptions of three types of technologies that can support community mobility: locator devices, community alert systems, and autonomous vehicles. We recognize that people also use public transit and specialized transportation systems for community mobility. We see locator devices, community alert systems, and autonomous vehicles as technologies that can complement and be combined with these systems.

12.1 LOCATOR DEVICES

Getting lost due to difficulties wayfinding in familiar and unfamiliar environments is a common concern among persons living with a cognitive impairment (Coughlan et al., 2018; Yatawara et al., 2017), and can be highly distressing for them and their families (Kwok et al., 2010). To ensure that this population can remain engaged within their communities for as long as possible, locator devices can be used to help compensate for their reduced wayfinding capabilities (Alsaqer and Hilton, 2015; Rassmus-Gröhn and Magnusson, 2014).

Figure 12.1: Locator devices. Shutterstock: Gergana Vlaykova.

Locator devices can be split into two main categories: (1) commercial locating devices and (2) mobile applications. For commercial locating devices, GPS, Radio Frequency Identification (RFID), and Bluetooth are the most commonly used types. GPS, similar to "X marks the spot" on a map, uses satellites to identify the position of the device. RFID, similar to the game "Marco Polo," transmits radio waves between a transponder, an antenna, and a receiver. The sound exhibited by the radio waves gets louder as the transponder gets closer to the receiver. Bluetooth, similar to GPS but predominantly used for indoor locations, relies on the close connection of other Bluetooth devices such as a smartphone to determine the location of the device (Alzheimer's Association, n.d.). Most commercial locating devices need to be purchased from an authorized dealer, and often require the user to purchase the locating device, in addition to paying for a monthly subscription fee which covers telecommunication costs. With the rise in the number of smartphone users over the years, mobile applications have become a common locating device for persons living with a cognitive impairment. They use existing features of a smartphone, such as GPS and Bluetooth, to identify its location from another device (i.e., another smartphone, computer, or tablet). Examples of such applications can include Life360 (https://www.life360.com/intl/), Find My Friend (https://apps. apple.com/us/app/find-my-friends/id466122094), and what3words (https://what3words.com/ products/what3words-app/).

As highlighted in the personas in Chapters 2 and 4, persons living with dementia, such as Michaela and Roger, use locator devices to remain connected to their community. Roger uses Life360 so family members can keep a lookout for him. This has been particularly helpful for Roger when he travels for various conferences and meetings on behalf of the Alzheimer Society of Canada and the Alzheimer Society of Alberta and Northwest Territories. In the event that he gets lost, his family members can help to guide him to his intended destination, despite them being hundreds of kilometers away. For other older adults living with a cognitive impairment, such as Betty, Apple Maps or Google Maps could assist by providing step-by-step directions on how she can get to the long-term care facility she volunteers at, or how to get to the library from her grandchildren's house when she is caring for them. She could also utilize a GPS to increase her sense of safety when using public transit.

12.2   COMMUNITY ALERT SYSTEMS

Whereas locator devices are strategies to be adopted by people, community alert systems are community-level strategies. They are aimed at creating a safety net in the environment rather than putting responsibility on the people whose safety may be compromised. Community alert systems utilize technologies, such as mobile apps and mass public messaging systems, to notify people in a particular geographic region about an emergency situation such as an impending storm or other safety threat but can also be implemented to address the safety concerns of a particular population, such as children with autism and persons living with dementia. Perhaps the most familiar community alert systems are Purple Alert and Silver Alert. These are outlined in Box 12.1.

Box 12.1: Purple Alert and Silver Alert. Author: Lauren McLennan (Research assistant, University of Waterloo)

Purple Alert

Purple Alert is a mobile app that can help locate a missing person living with dementia (Alzheimer Scotland, 2018; Dementia Circle, 2017). It is offered, monitored, and supported by Alzheimer Scotland (https://www.alzscot.org/purplealert/FAQ).

The app supports efforts to find a missing person living with dementia (Alzheimer Scotland, 2021). It hosts an online community of persons living with dementia, caregivers, family members, friends, and community members (Alzheimer Scotland, 2019). Persons living with dementia and/or their caregivers can upload information about the person living with dementia to the app, including identifying characteristics and an up-to-date photo (Alzheimer Scotland, 2021b; Dementia Circle, 2017). Caregivers can release their loved ones' profile information if the person is missing. App members receive a notification of the missing person if they are within 30 miles of the last known location of the missing person (Alzheimer Scotland, 2018; Dementia Circle, 2017). Members can then keep an eye out for the missing person, acting as additional eyes and ears for police (Alzheimer Scotland, 2019). Once a missing person is found, the alert is discontinued, and app members are notified (Dementia Circle, 2017). If a missing person is found by an app member, the member can contact the missing person's caregiver via the app, which allows the person living with dementia and their caregiver to be reconnected. Purple Alert is designed to support locating a missing person living with dementia, but it does not replace emergency services. Police should still be contacted as soon as it is known that a person is missing (Alzheimer Scotland, 2019).

The Purple Alert app is free to download and reached 10,000 downloads in late 2019 (Alzheimer Scotland, 2019; Dementia Circle, 2017). Since the app became available in 2017, 13 missing persons were found by community members who used the app (Alzheimer Scotland, 2019). The app was re-designed and re-launched in 2019 to incorporate user feedback and missing incident data to better suit community needs (Alzheimer Scotland, 2018, 2019).

Silver Alert

Silver Alert, an alert system for missing older adults used in the United States, also uses the public's assistance to find missing persons. Modeled after Amber Alert, a Silver Alert is activated once an older adult is reported missing to police (Fernandes and Colello, 2009). A Silver Alert can be broadcast using media channels such as television, radio, and digital highway signs (Fernandes and Colello, 2009). The public are asked to report any information regarding the missing person (Fernandes and Colello, 2009). Once the person is found, the Silver Alert is discontinued (Fernandes and Colello, 2009).

As of 2017, all but a few states implemented a Silver Alert program (Gergerich and Davis, 2017). Literature on the efficacy of Silver Alert is inconclusive (Fernandes and Colello, 2009). Given the differences between state Silver Alert programs (i.e., activation criteria, activation duration, dissemination channels, reporting, and program evaluation), it is difficult to determine its overall effectiveness.

Another community alert system that targets the community at large is Safeland (Safeland, 2021). Safeland is used in Sweden and some parts of the United Kingdom as a means of helping people in a neighborhood or geographic location communicate and collaborate with one another. It is available as a mobile app and also as a web app. Safeland assists members of a particular community to share information and resources and helps them to work together to improve safety. A member can ask other members to check in on an individual or to keep a lookout for individuals who may be lost. It can also be used to report crime and create safer communities and in this way, Safeland is also a virtual neighborhood watch program.

A variation of a community alert system is the Mimamoriai service (Mimamoriai Project, 2021) which has been implemented in Japan. This service combines an SOS sticker system with a mobile app. SOS stickers contain an emergency phone number and an identification number and can be placed on the belongings that are carried by people who are at risk of getting lost or experiencing emergencies while in their communities, such as persons living with dementia, children, and people with medical conditions. These belongings can include wallets, keys, umbrellas, bags, hats, coats, and so on. If people require assistance, bystanders call the emergency number on the

SOS sticker which connects them by phone to family members but without disclosing the family member's phone number. The sticker system can be combined with a mobile app that allows those individuals within a particular region who are members of the group to communicate with one another. Members can request the help of other members to locate loved ones by sending a picture, a physical description of their loved one, and other information that may help locate the lost person. Alternatively, members can decide not to send a picture and detailed information about their loved one and instead only provide the identification number contained on the SOS sticker. They can also select the search range (radius of 500 meters to 20 kilometers). As of early 2021, the Mimamoriai system has been downloaded 800,000 times and has been adopted by local governments in Japan (Aging 2.0, 2021). Mimamoriai can be a particularly useful tool for those who utilize public transit or specialized transportation systems.

12.3   AUTONOMOUS VEHICLES

Another technology that can support independence in community mobility, and thus can enable community participation, is autonomous vehicles. While this emerging technology has not been fully developed nor implemented on a widespread basis, it is anticipated that autonomous vehicles will become a reality in the future.

Autonomous vehicles, also known as driverless, robotic, and self-driving cars (Fleetwood, 2017), are vehicles in which parts or all of their operations are removed from being controlled by the driver and are instead controlled by the vehicle itself (Faber and Lierop, 2020). Sensors are installed within and on the outside of autonomous vehicles to perceive and monitor the driver, the environment through which the car is traveling (e.g., traffic lights, curbs, pedestrians, and other vehicles), and the speed and motion of the car itself (Van Brummelen et al., 2018). Data collected through these sensors are then analyzed and an algorithm is followed to automate the vehicle's response in accordance with the data collected. The Society of Automotive Engineers developed five levels of automation that applies to autonomous vehicles. These are outlined in Figure 12.2 and overlaid with examples of how they can apply to driving.

Semiautonomous and autonomous vehicles have been suggested for people who may not possess the complex skills and abilities to safely operate a car due to the presence of conditions that affect their physical and cognitive abilities (Rogers et al., 2020). In-vehicle technologies such as warnings (lane departure, forward collision, blind spot), parking assistance, navigation, adaptive cruise control, and adaptive headlights could benefit older drivers and thus may improve their safety, driving ease, and comfort (Eby et al., 2016). In particular, those who experience declines in attention, processing speed, ability to scan the environment, and memory could benefit from some of these technologies found in semiautonomous vehicles such as intersection identification and front/rear cross-check systems, lane departure and blind spot warnings, and navigation (Knoefel et al.,

2019). These could reduce common driver errors, such as unsafe turns, drifting into another lane, inaccurately assessing distances/gaps between vehicles, inability to quickly responding to what is happening around them, and not seeing or responding to traffic signal or signs (Classen et al., 2010; Regan et al., 2011). Semiautonomous and autonomous vehicles may also pose a solution for those who are capable of driving but have driving anxiety (Taylor et al., 2018).

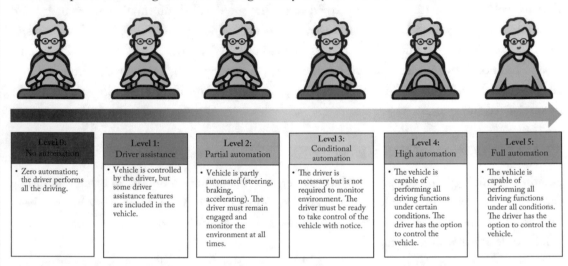

Figure 12.2: Levels of automation. Courtesy of Hector Perez. Adapted from: (1) National Highway Traffic Safety Administration. (2017, September 7). NHTSA. NHTSA; U.S. Department of Transportation. https://www.nhtsa.gov/technology-innovation/automated-vehicles-safety; (2) SAE International. (2018). *Taxonomy and Definitions for Terms Related to Driving Automation Systems for On-Road Motor Vehicles* (SAE International Standard). https://doi.org/10.4271/J3016_201806; and (3) Fleetwood, J. (2017). Public health, ethics, and autonomous vehicles. *American Journal of Public Health,* 107(4), 532-537.

Semiautonomous in-vehicle technologies are already available in some models from Ford, Tesla, Mercedes, Toyota, Honda, Cadillac, Volvo, and others. Some companies are also implementing and testing vehicles containing higher levels of automation. Voyage (https://voyage.auto/about/) provides on-demand taxi services to older adults in two retirement villages in California and Florida (https://voyage.auto/communities/). Serving close to 130,000 older adults, Voyage robotaxis have been deployed on close to 800 miles of roads within these retirement communities, taking the residents to within-community appointments and activities. Waymo One (https://waymo.com/waymo-one/), which began as the Google car, is an autonomous ride-hailing service using level 4 autonomy. Although it is not deployed specifically for older adults, it is being used in less densely populated test communities on public roadways in Arizona. Waymo One will begin testing in communities that experience winter weather conditions.

Even if semiautonomous and autonomous vehicle technologies are available, they must be accepted and adopted by people to make a positive impact on their lives. People are open to adopting semiautonomous vehicles if their safety is proven, costs of the vehicle, insurance and repairs are affordable, and the conditions under which they best work are clear (Robertson et al., 2018). However, older adults are more hesitant about using fully autonomous vehicles (Kadylak et al., 2020; Rogers et al., 2020). Nonetheless, the flexibility of on-demand transportation to overcome older adults' existing challenges with community mobility makes autonomous vehicles an attractive option for some older adults (Faber and Lierop, 2020).

Many ethical questions have yet to be asked and resolved pertaining to autonomous vehicles including decision making and moral dilemmas in extreme traffic situations, accountability and responsibility, privacy, and sustainability (Collingwood, 2017; Goodall, 2016; Hevelke and Nida-Rümelin, 2014; Lundgren, 2020; Taiebat et al., 2018). These ethical considerations must be considered in the algorithms inherent in autonomous vehicles (Fleetwood, 2017). Fully autonomous vehicles that are widely rolled out in communities with different population densities and weather conditions are still many years away. In the meantime, older (and younger) adults will rely on semiautonomous technologies and other modes of transportation, including public transit and special transportation systems, to support their independence in community activities.

FIND OUT MORE

1. **Purple Alert:** https://www.alzscot.org/purplealert

CHAPTER 13

Usability of Technologies to Support Independence

WHAT IS IN THIS CHAPTER?

This chapter describes the concepts of the acceptance, adoption, and usability of technological and health care interventions. We also explain the factors that facilitate and the barriers that hinder technology acceptance, adoption, and use by older adults. This chapter includes a Box in which we present an example of how to determine the technology acceptance and usability of Global Positioning Systems for people with dementia and their caregivers. The latter part of this chapter is dedicated to the concept of commercialization and factors than hinder the commercialization of technologies.

PERSONA, SCENARIO, AND SOLUTION

Enma Cruz Delgado

- **Age:** 72
- **Gender:** Woman, pronouns are she/her
- **Lives in:** Single family condo with her daughter, in Miami, Florida, USA
- **Social circle:** Family, a few close friends, older adults in a senior community program

Signature interests
Crafts such as two-needled and crochet-knitted, and sewing. She is a self-learner and loves to read.

" I try to learn everything I can and I use technology for this. I do it because I like to feel that I am capable of doing something that I never in my life thought I was going to do, and it pleases me."

Health
Blood pressure, overweight, hypothyroidism, and prediabetes.

Technology
Uses a smartphone, a laptop computer, and a smartwatch. Has internet connection and regularly uses social networks and streaming platforms. In addition, she has smart home devices such as a personal voice assistant, a smart thermostat, and a smart doorbell.

Persona. Enma is a 72-year-old woman who was born in Cuba. She would like to have studied microbiology, however when she finished high school she could not attend university. At that time, some families believed that a woman's role was to be a wife and a mother rather than a professional. She married and had a son who lives in Canada, and a daughter who lives in the United States. She was widowed when she was 40 years old and never re-married. She emigrated to Miami in 2017 at the age of 68, and since then she lives with her daughter in a condo. Her daughter is a health care provider who has long shifts at the hospital, so Enma spends many hours alone at home. Enma has some medical conditions such as high blood pressure, obesity, hypothyroidism, and prediabetes, although she is independent in her activities of daily living. Three times a week she attends a senior community program delivered in her mother tongue (Spanish). In this program, she interacts with other Cubans who also emigrated to the United States as older adults.

Scenario. Enma lived most of her life in Cuba, where there was no access to the Internet or devices such as smartphones. Enma got access to these technologies only recently when she immigrated to the United States. She felt a new world opened before her eyes when she gained access to everyday digital technologies and perceives that the technologies give her opportunities for endless learning. As she did not realize her desire to attend university, she now uses the technologies to learn as much as possible.

She uses technologies primarily for learning, maintaining relationships with family and friends, for entertainment, and to facilitate self care. Enma uses the laptop for informal lifelong learning. For example, she watches a wide range of educational videos on YouTube, ranging from how the human body and the cells work to sewing techniques. She also uses the laptop and the smartphone to play video games regularly and talk with her family in Canada and in Cuba, which only recently was connected to the Internet.

In relation to self care, she wears a smartwatch all day, even to sleep, to monitor her blood pressure, heart rate, weight, and the number of steps she takes. She uses this information to change her health behaviors, i.e., to increase her daily physical activity. Enma uses her smartphone to make medical appointments and organize transportation service to the clinic. She appreciates that she can do these health management activities independently using technology without asking her children or clinic staff for help.

Enma started using everyday digital technologies when she was almost 70 years old. She thought that it was difficult to use technology for the first time, but finds it gets easier with practice. She has learned to use functions to enhance her experience with technologies, for example, zoom in to enlarge font size. She perceives technologies to be intuitive, so she doesn't feel like she has to put much effort into using them. Also, if something is not easy, she perseveres until she succeeds.

Solution. There is a common ageist belief that older adults cannot learn how to use high tech. Enma's relationship with technology challenges this belief. She enjoys using technology and has been able to learn despite accessing digital technologies as an older adult. With the use of the technologies, Enma sees herself as a more educated person. For example, before she had access to technology, she did not dare to speak when she was in front of a professional, or other more educated peers. Now, she participates confidently in conversations with them.

Enma is aware that she cannot learn as fast as when she was younger, but access to technologies has given her an opportunity to learn new things. Thus, she has gained confidence to talk to people. She also feels that although her encounter with technology occurred later in life, she has found in technology the opportunity to participate in activities that are important, meaningful, safe, and fun which contributes to personal fulfillment.

13.1   ACCEPTANCE, ADOPTION, AND USABILITY OF TECHNOLOGIES

The ultimate goal of developing technology is for it to be used by its end users. Although this may seem like a trivial statement, it is not. From a technology standpoint, studies have shown that the rate of abandonment is high and, as a result, adoption is low (Petrie, H., Carmien, S., and Lewis, A. W., 2018). We have posited that the acceptance of technologies precedes adoption and, in turn, adoption precedes usability (Liu et al., 2019). Whether technology is used or not can also inform the technology acceptance process.

13.1.1   WHAT IS TECHNOLOGY ACCEPTANCE?

Technology acceptance relates to users' beliefs. Technology acceptance occurs when a system of individuals' beliefs provokes a behavioral change toward deciding on using a given technology. These beliefs include that technology helps a person to achieve one's goals, is easy to use, is enjoyable to use, the benefits of using the technology are worth the monetary cost and significant others believe the person should use the technology (Venkatesh et al., 2012). Simply put, technology is accepted when the user perceives it is good enough to satisfy one's needs (Nielsen, 1993). For example, Enma believes that using the smartwatch helps her (it is useful) to improve her overall health condition (her goal). She also believes the smartwatch is easy to use as she found it intuitive. The cost is reasonable considering the benefits she believes the technology is bringing to her. She also believes that her children appreciate her using the smartwatch to track her blood pressure, daily physical activity, and weight (others' influence on Enma's beliefs).

13.1.2   WHAT IS TECHNOLOGY ADOPTION?

Technology adoption deals with the change of acceptability over time. Technology adoption refers to the time it takes for a group of individuals (the adopters) to incorporate new technology into their lives consistently. The technology adoption by each individual of a social group follows a process of five steps: (1) the individual becomes aware of a new technology existence (awareness), (2) the individual finds out enough information about the technology's features (persuasion), (3) the individual makes the decision whether to adopt or reject the technology (decision making process), (4) the individual acts accordingly on his or her decision (implementation), and (5) the individual engages in an ongoing assessment process on whether to continue or discontinue with the technology adoption decision (confirmation) (Rogers, 1995) cited by Straub (2009). In addition, other factors influence the adoption of technologies such as the political, economic, regulatory, and sociocultural context as well as their successful commercialization (Greenhalgh et al., 2017). While living in Cuba, Enma was aware of the existence of everyday digital technologies and the Internet; however, she could not move toward the adoption of these technologies, as in the context she lived in, she had no access to them. When she moved to Miami, and her context changed, Enma learned more about these technologies and decided to adopt them.

13.1.3 WHAT IS USABILITY?

Usability is a concept associated with the actual use of technology. In other words, usability is the user experience when interacting with technology. Usability is a multidimensional concept with five attributes, namely: learnability (easy to learn the technology), efficiency (level of productivity achieved with minimum required resources), memorability (easy to remember how to use the technology), free of errors (free of error or low error rate), and satisfaction (pleasant to use) (Nielsen, 1993). Designing technology based on universal design principles (see Chapter 2) facilitates its use, especially by some populations such as older adults. For example, a novice older adult aged 70 or older consumes twice the time to learn to use an ICT-based system compared to a novice younger adult in the 20s (Charness et al., 2001). Enma found that some technologies she uses are intuitive and straightforward; as a result, her experience using them (e.g., personal voice assistants such as Alexa) is positive and pleasant. However, for other technologies such as the smart watch, she has to put more effort into using them (e.g., synchronizing the smartwatch with the cellphone). If Enma found the smartwatch extremely hard to use, she would probably abandon it.

13.2 HOW TECHNOLOGIES ARE ACCEPTED, ADOPTED, AND USED BY OLDER ADULTS

AgeTech such as electronic assistive products and everyday digital technologies can support older adults' independence. However, barriers to acceptance, adoption and use of these technologies by older adults include stigma, contextual barriers, and inappropriate or unrealistic designs.

Stigma can hinder the acceptance of high tech. In general, older adults like to feel independent; thus, if high tech is presented to an older adult as an assistive product, the technology may conflict with the older adult's identity as an independent person, it can evoke stigma associated with disability and weakness and generate rejection or at least ambivalence, ending up discouraging acceptance (Jovanović et al., 2021).

Contextual barriers hinder the adoption of high tech. Some political, institutional, and economic contexts in which an older adult lives do not have the conditions for technology adoption

(Hunter et al., 2021). One example is disparities in internet access in some geographical areas. This is a reality in developing countries where the bandwidth is limited and in some areas of industrialized countries. For example, in 2017, 42% of New York City older adults lacked broadband internet access, compared to only 23% of people aged 18–64 (Eberly et al., 2020). Since many high techs rely on an internet connection, where the Internet connection cannot support high tech, their adoption simply cannot occur.

A design of technologies that does not meet older adults' needs and preferences also affects acceptance, adoption, and use (Allemann and Poli, 2020). A problem for the adoption of high tech may be that while developers are focused on the technology providing assistance to the older adult, older adults emphasize a pleasant experience of using them (Jovanović et al., 2021). For example, in the case of wearables, despite the fact that they are perceived as useful by older adults (Kekade et al., 2018), problems that affect their use include sweating, itching, and discomfort, which interfere with the sensitivity of the monitoring system (Cruz-Sandoval et al., 2021). This can significantly affect the adoption of these technologies; a study found that 1 in 3 older adults abandon wearable cardiac monitoring technologies within two weeks after purchase (Ferguson et al., 2021). Regarding smart homes, studies assessing the usability of these technologies are usually focused on the technical aspects of sensor use with older adults or tested in controlled environments which do not represent the reality of older adults' living situations (Hunter and Lockhart, 2020). As a result, these technologies may not be adopted in real-world scenarios.

Solutions to improve the technology acceptance, adoption, and use of high tech by older adults include, whenever possible, showcasing the technologies as AgeTech or everyday digital technologies rather than as assistive technologies, which can reduce stigma, and to involve older adults in all steps of the design of technologies through co-design and co-creations approaches (Knight-Davidson et al., 2020). Furthermore, if designers conceive the new technologies outside the health care service, they may create technologies that support older adults aging in place. For example, when smart homes and telerehabilitation systems have allowed older adults to customize the communication channels, including members of the social network rather than the traditional patient-to-provider channel, acceptance has increased (Hunter et al., 2021). Implementation of principles of universal design and co-design involving older adults from the early stages of the development process of new technology can also increase the use and reduce abandonment. Addressing these issues would challenge stereotypes that older adults are disinterested and incapable of using high tech (Sinner and Wei Lim, 2021).

Box 13.1: Determining the technology acceptance and usability of Global Positioning System for people with dementia and their caregivers

The GPS technology was tested with 45 dyads (people living with dementia and their caregivers) that used the devices for an average of 5.8 months over a 1-year period. They lived in community settings in rural and in urban areas. Types of GPS devices were a simplified cell phone that could be worn on a lanyard or belt, a pair of insoles, and a watch (Liu et al., 2018, p. 638). In this study, a combination of methods was used: (1) pre- and post- paper-based quantitative questionnaires to understand what factors affected the use of the GPS devices; (2) seven focus groups; and (3) logging actual use, i.e., "client-caregiver dyads were contacted on a weekly basis by phone to respond to a structured questionnaire based on their experiences using the GPS" (Liu et al., 2017). The study concluded that "the GPS provided caregivers with peace of mind and reduced anxiety in dyads when their clients became lost," and that "the dyads would continue to use the GPS devices if they were able to do so" (Liu et al., 2017).

13.3   COMMERCIALIZATION OF TECHNOLOGIES

If usability is associated with the actual use of technology, then usability can be determined only when technologies are commercially available. **Commercialization** is the process of introducing new products into commerce for people to purchase and use (Mihailidis and Sixsmith, 2021). The development of technologies involves a number of steps, namely, ideation, invention, market research, intellectual property research, defining the technology and the key features of interest, developing the technology, testing the technology, and finally, implementation and commercialization (Mehta, 2008, p. 11). Commercialization is the final step in a number of phases that a new product has to go through before it is available on the market.

The commercialization of technologies can be hindered by several factors. Innovation itself and its level of technological maturity, the adopters, and the implementation process (e.g., dedicated monetary resources, public support for commercialization, and innovation procurement) are key factors (Greenhalgh et al., 2004). The level of technological maturity, also known as **technology readiness level** (TRL) (Mankins, 1995) is an indicator that measures how products move from initial concept to their final implementation. At first glance, one might think that the more technologically mature a product is, the more the commercialization of this technology will be achieved. Unfortunately, increasing technological maturity does not translate to a successful commercial product (Héder, 2017). Other aspects such as disruptive competition, branding, the purchasing power of the end users, and financial support during the commercialization phase should be taken into account. Technology usability, acceptance, and adoption are understood to be complex systems of belief and processes that adopters experience and relates to whether a technology has social and practical acceptability (Liu et al., 2019). Studies have shown that not involving end users during the early stages of technological development (obtaining user requirements) in a collaborative manner, known as user-centered design (UCD) or co-design (Mihailidis et al., 2011), is an error that affects adoption and commercialization (MacNeil et al., 2019). However, the so-called **Valley of Death** (see Figure 13.1), a metaphor used to describe the gap that exists between research and development, and the commercialization of a new technology, is a key challenge for technologies and innovations (Collins et al., 2016).

The Valley of Death has become a wide and deep gap for many health care innovations, funding issues being the main cause. For many health innovations, the Valley of Death occurs because the developer or researcher of the technology has demonstrated its efficacy, but there is a lack of financing for the implementation, scale up, and manufacturing processes. At this point, the researcher is trapped in a "blind alley," or there is no funding or investment, as a government considers the technology to be too "applied" to continue to provide funding, yet the private sector does not want to invest capital because the technology has not yet been implemented (Frank et al., 1996). A recent study demonstrated that for universities, research and development spending is more focused on basic research, with 75.8% of the total fund (Ferguson, 2014). The end result of this pattern of support in research and development is that innovations do not translate into actual applications (Héder, 2017). For example, a recent report showed that more than 90% of all university inventions fail to be adopted by industry (AUTM, 2011).

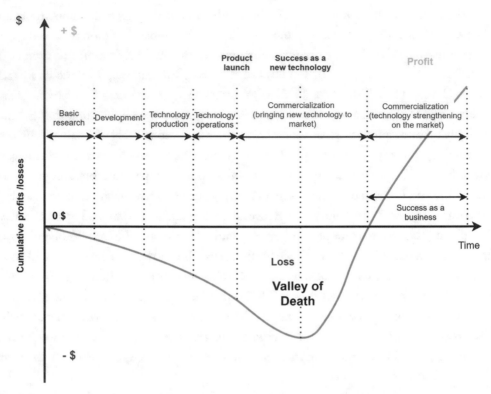

Figure 13.1: The Valley of Death. Modified from (Osawa and Miyazaki, 2006).

Commercialization using public procurement or a combination of private and public procurement is a solution to bridge the Valley of Death. However, it is no less challenging for innovation companies. Institutional bias, higher costs (e.g., administrative overheads), regulatory burdens (e.g., procurement rules and practices), and uncertain market size have been cited as the main factors that make it difficult to find private sources of funding to further develop innovative products and companies (National Research Council, 2007). The application of the Public Procurement of Innovative solutions can be used in cases where innovations are not yet available on a large-scale commercial basis (European Commission Directorate General Environment, 2009; OECD, 2017). There are several ways in which a Public Procurement of Innovative solutions can influence the course of an innovation. Another option is to have a combination of public and private procurement. For example, the Small Business Innovation Research program is one of the largest examples of United States public-private partnerships (National Research Council, 2007). For Canada, in particular, the federal government has created the Office of Small and Medium Enterprises (OSME) in Public Services and Procurement Canada (PSPC) to advocate on behalf of small and medium-sized enterprises (SMEs).

Box 13.2: The Innovation–Commercialization decision box

The decision to stop investing in developing new technology can be challenging. There is no easy answer to this decision-making process. The Innovation–Commercialization traffic light model may help you to make this decision. Draw two axes that measure the level of technology and commercialization readiness of your invention/technology/product. Then position your invention/technology/product in one of the quadrants depicted in the figure. In the quadrants where the technology readiness level is high, keep going! You either need more money or to expand your business. In the quadrants where the technology readiness level is low, be careful, whereas if the commercialization readiness is potentially high, hurry up as there is a gap in the market and a business opportunity. On the contrary, if the commercialization readiness is definitely low, stop the project or try something else!

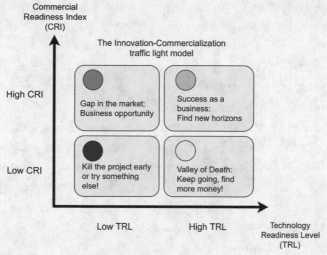

Note: Figure elaborated by the authors of this book.

If inadequate evidence hinders commercialization, perhaps one way of solving this problem is the application of more **implementation research**, which is defined as "the scientific study of methods to promote the systematic uptake of research findings and other evidence-based practices into routine practice, and, hence, to improve the quality and effectiveness of health services" (Eccles and Mittman, 2006, p. 1). Of course, implementation science is not a panacea. Barriers to engaging in implementation research such as knowledge and awareness of implementation science, attitudes about this type of research, the career benefits of implementation science, research community support, and research leadership support persist (Stevens et al., 2020). Therefore, more funds, diffusion, and dissemination should be provided to implementation research for it to gain attention from the scientific community.

A new wave of entrepreneurship is emerging, the **seniorpreneurs**. Seniorpreneurs are older adults who decide to work beyond the traditional retirement age. As a new career path, they start businesses to address needs of other older people they can relate to. It is not just their money assets but their social and relational capital, and their experience and wisdom to mentor entrepreneurs, that create a positive cycle of entrepreneurship, that can lead to revenue, generate new market opportunities, and foster innovation in AgeTech. There are challenges and opportunities in this area, the most important being training. Seniorpreneurs with bold ideas deserve mentorship, training, and resources to deliver their vision. This is the case with Jamie, our persona in Chapter 5. As a retired police officer who has started a technology company, he has experienced challenges in finding investors.

Box 13.3: Seniorpreneurship, a boosting trend

The term "senior entrepreneurship" or "seniorpreneurship" has been defined as "people aged 50-plus starting up in business" (Kautonen et al., 2008). Although it is a fact that older people are less likely to engage in entrepreneurial activity than younger people, this trend is changing: seniorpreneurship is increasing. The literature reports three main reasons why there is an increased interest in senior entrepreneurship. First, as populations age, the number of older business founders will also increase. For example, in Canada, nearly 15% of Canadians 65 years and older are still in the workforce and their entrepreneurial spirit is strong. Second, governments are releasing policies and programs intended to reduce older-age unemployment, increase the social inclusion of older individuals, and to enhance the innovative capacity of the economy by employing the human and social capital of mature individuals through new start-ups (Botham and Graves, 2009). Third, people aged 50-plus still have "personal resources" such as self-efficacy and an accumulated experience that they are willing to "send back" to the community and to put to the service of others.

Reduced motivation has been found as the main factor that hinders seniorpreneurship. Therefore, some motivation-oriented measures (e.g., programs/policies) have been created in order to increase older adults' inclination to pursue entrepreneurial activity. For example, removal of ageist bias in senior entrepreneurship, programs with specific target groups such as older women, and efforts to increase older people's awareness of entrepreneurship as an option for a "second" late career. Example of policies/programs are: PRIME – The Prince's Initiative for Mature Enterprise (UK), SCORE (USA), Best Agers (Denmark, Estonia, Germany, Latvia, Lithuania, Poland, Sweden and the UK), and New Enterprise Incentive Scheme (Australia).

Box 13.4: Seniorpreneurship, the LaceUp™ case

LaceUp™ is an Alberta start-up that is 50–50 co-owned by Drader Manufacturing Industries Ltd and Alfego Corp. The partners, Mr. Gord McTavish (Drader) and Mr. Bryan Goutrho (Alfego), are two senior entrepreneurs. They have a passion to create and market products that enhance the quality of life for aging Canadians who want to remain active, like themselves. While their product, LaceUp™ is designed for use by people across the lifespan, it is particularly appropriate for use by older adults who have a diminished range of movement in their spine and lower extremities which make reaching down to the feet challenging. Inspired by their own active lifestyles and the daily activity of lacing up their running shoes, the creators of LaceUp™ collaborated with the University of Alberta's Faculty of Engineering to design the foot stool that can be used by individuals in a seated or standing position. They tested the usability of the LaceUp™ device (in partnership with researchers at the Department of Occupational Therapy at University of Alberta) with health professionals. The results of the study are now being used as evidence to scale up the applications and benefits of this product to other populations such as health care workers, young adults with chronic conditions including obesity and arthritis, skaters who require a tool to stabilize their skates while tying the laces, and foot care activities in diabetic populations.

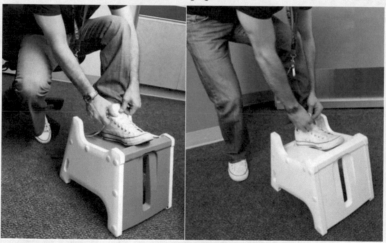

FIND OUT MORE

1. World Leaders in Research-Based User Experience, Nielsen Norman Group: https://www.nngroup.com/

2. Usability.gov, Improving the User Experience (UX): https://www.usability.gov/

3. Senior entrepreneurship: https://www.usability.gov/

4. CIMIT's Guidance and Impact Tracking System (GAITS): https://www.gaits.org/ja/

5. Healthcare Innovation Cycle for "MedTech" Solutions: https://www.gaits.org/ja/web/medtech/guidance

PART IV

Challenges and Future Directions

CHAPTER 14

AgeTech for Autonomy and Independence: Challenges and Future Directions

WHAT IS IN THIS CHAPTER?

In this last chapter, we highlight challenges and future directions related to the use of AgeTech to support autonomy and independence in older adults. These challenges pertain to:

- the meaning of autonomy and independence

- society inequities and the digital divide

- evolution of technologies

- the importance of policies to create and sustain changes

Through the persona of three older adults—an Indigenous Elder in Canada, an older adult in West Africa, and a frail older woman in Japan—we see the opportunities for autonomy and independence facilitated by technology. But such AgeTech solutions are beyond reach for those without family members, communities, or resources to bridge the digital divide. We conclude with suggested future directions to address these challenges.

PERSONA, SCENARIO, AND SOLUTION

Authors: Nicole Bird and Susie O'Shea

Leslie (Les) Nelson

- **Age:** 66
- **Gender:** Man, pronouns are he/him
- **Lives in:** Single-person unit—lives alone, in Vancouver, British Columbia
- **Social circle:** A few close friends and Carnegie Community Centre patrons have become his family

Signature interests

Watching sports on television, learning about new technology (new smartphone, new laptop), social media, and singing and drumming.

" Everyday is a lifelong learning opportunity even after graduation from school. Technology is my learning now, and I feel I am always asking questions to become more experienced."

Health

Diabetes, high blood pressure, anxiety, arthritis, just found out has a cataract in left eye.

Technology

Has a smartphone, tablet, and laptop computer. Has internet connection at home and the Carnegie Community Centre that he regularly uses to visit social networks and social media platforms.

Persona. Elder Les is a 66-year-old man who was born in Bella Coola (Nuxalk Nation). Les is Kwakwaka'wakw, Gusgi'mukw, and Nuxalk. He came to the Downtown Eastside of Vancouver in 1972 and fell into the "darkside" until 1995 when Les contracted tuberculosis (TB). Les started to walk the "Red Road" in 1995 toward an opportunity to go back to school to earn his high school diploma. Les continued to struggle with pain which was an effect of the TB that brought on arthritis in his right wrist and lower spine. He persevered through life and became a volunteer at the Carnegie Community Centre in 2004. Les still volunteers at the Learning Centre where he works as a receptionist. Although the Learning Centre has been closed due to COVID, Les looks forward to the reopening in the Fall of 2021. Since 2016, Les is the Elder in Residence at the Carnegie Community Centre, with a ceremony acknowledging his role with community members and representatives from the xwməθkwəy̓əm (Musqueam), Skwxwú7mesh (Squamish), and Səl̓íl-wətaʔ (Tsleil-Waututh) Nations. As the Elder in Residence, Les has taken on the responsibilities of cleansing the community centre monthly through smudging, hosting a weekly Elder Chat program where community members drop in to visit, Cultural Sharing Program activity lead, ceremonial acknowledgements, and providing Indigenous leadership to center staff.

Scenario. During the past few years, Les started to connect with various forms of technology using a basic smartphone and later a tablet for social media and games. Learning how to use both pieces of technology did not come easy to Les; he was very intimidated by the phone and sought out community technology classes to support his learning. This was a whole new world of having information at his fingertips and being able to ask a question to Google and having an answer instantly. Les started using his smartphone to text message or call to connect with family in Quatsino and the Lower Mainland along with being able to make medical appointments to keep his health and well-being looked after. Les also uses his smartphone to access a Kwak̓wala language learning app and watches videos of Bighouse singing and dancing on YouTube. The smartphone also enables Les to connect with the Indigenous Programs Coordinator (Nicole Bird) at the Carnegie Centre to confirm his weekly schedule and receive reminder texts and phone calls for upcoming bookings.

Les was feeling very comfortable with basic technology and working as the Elder in Residence at the center. In March 2020, the COVID-19 pandemic resulted in the cancellation of all in-person programming and ceremonies due to safety restrictions. The community was left with no access to Indigenous programs that provided them a cultural connection and Elder support. Les also lost his connection to the community and was also not permitted to receive visitors in his place of residence. Les had to go and search for Wi-Fi in local shops, where he would stand outside the stores to use his smartphone to access social media.

"I thought this was only going to last two weeks. The days became months and then the months became a year. In the snap of a finger, we have been living with COVID for a year plus."

Solution. Les was asked to consider ways that he could connect with the community in a "safe way" during COVID. Les started to do an outdoor "Cultural Sharing to-go" program which enabled him to connect with community members in a safe way. Les was able to provide sacred medicines, cultural crafts, smudging, and Elder support. This "to-go" approach was also used to develop an Elder-in-Residence online program. Les identified opportunities to use technology to connect with others in the community and beyond. Les and Nicole set up a studio to do a weekly online livestream "Elder Chat" program to share songs and stories on the "Cultural Sharing Program" Facebook group. Les worked with Susie O'Shea of the University of British Columbia Learning Exchange to share recordings of his Elder Programs in the Learning Exchange's "V6Activities" YouTube channel. Les created a "Virtual Smudge" ceremony where viewers were taught how to smudge and receive the benefits of sage medicine through the computer screen. Les talked to viewers as if they were with him in the room and shared the importance of connecting to ceremony and culture in a virtual way. Les created cultural packages that could be mailed to viewers. These included prayer ties, sacred medicines, coloring sheets, crafts, and masks that were sewn by Elders at the Carnegie Centre. In his role as the Elder-in-Residence, Les was also able to participate in virtual celebrations such as "Hearts Beat" and "National Indigenous Peoples Day." Les participated as a singer and lead of the drum group lexwstilem with the "Heart of the City Festival." As he became more comfortable with online programming, Les moved forward with installing Wi-Fi in his own home. He used his new laptop to access more online media and gained experience with Zoom and Webex meetings. With laptop access, Les became part of committees, attended workshops, and offered elder support to outside agencies. Les has continued to use technology to access medical appointments, family and friends and continue to offer programs to connect with the community. The past year has been a learning experience for Les that has moved him from using a basic smartphone to feeling confident and comfortable with using a tablet, laptop, and more advanced smartphone. Les has gained new friends, connections, and opportunities through his technology journey. With other Indigenous Elders, he provided ceremonies and support online using everyday digital technologies.

14.1   FROM AUTONOMY AND INDEPENDENCE TO RELATIONAL AUTONOMY AND INTERDEPENDENCE

We saw in Part I (Chapters 1 and 2) that the dominant Western view of autonomy and independence does not reflect the reality of our older adult populations. Autonomy is related to our sense of identity. According to Dworkin (2015, p. 14), "… autonomy is conceived of as a second-order capacity of persons to reflect critically upon their first-order preferences, desires, wishes and so forth and the capacity to accept or attempt to change these in light of higher-order preferences and values. By exercising such a capacity, persons define their nature, give meaning and coherence to their lives, and take responsibility for the kind of person they are." It is generally accepted that people value autonomy and this aspect of having a choice must be respected. However, the choices individuals make are influenced by their development, self-evaluations, and social relationships (Entwistle et al., 2010). Therefore, service providers would support older adults' behavioural goals if they take a perspective of relational autonomy, in other words, what goals older adults desire to achieve given the influence of their social networks.

14.1.1   RELATIONAL AUTONOMY

Relational autonomy can be seen in the example of the use of technologies in two studies. In the GPS study conducted by Liu et al. (2018), there was a positive association on the "social influence" responses to adoption of GPS devices between participants living with dementia and their caregivers. There was "positive feedback" that "mutually influenced" the intention of caregivers and persons with dementia to use the devices in order to alleviate caregivers' concerns (2018, p. 642). This was validated by comments including "*using the devices to help with family support, less family stress or to find a way to have easy communication with my wife and family*" (2018, p. 642). In contrast, social relationship was not an influence in another study that examined factors for adoption of rehabilitation technologies among therapists (Liu et al., 2015). In other words, therapists were not influenced by social pressure from peers nor supervisors to use innovative technologies in their practice setting. Instead, therapists' choice to use new technologies was influenced by *performance expectancy* of the technologies to make a difference in helping their clients and by institutional support. In the first study, people's intention to use the GPS devices was influenced by consideration of each other's well-being. In the second study, the decision to use rehabilitation technologies was not influenced by pressure from peers or colleagues, rather it was based on a user's ability to help a client.

14.1.2 INTERDEPENDENCE

Like our tendency to frame autonomy out of context of an older adult's social influence, we tend to view independence out of context of an individual's relationship with one's environment. As presented in Chapter 1, we are all interdependent in human and non-human aspects of our lives. This is most evident in Indigenous worldviews. The Canadian Model of Occupational Performance and Engagement (CMOP-E) (Polatajko et al., 2013) describes the relationships between an individual's spiritual, affective, physical, and cognitive functions with activities of self-care, productivity, and leisure in the context of environments (physical, institutional, cultural, and social). In 2019, Fijal and Beagan introduced a version that focuses on Indigenous ways of knowing (Box 14.1). This is an example of how concepts of autonomy and independence are adjusted to reflect the intersections of race, culture, gender, age, dis/ability, among other aspects of older adults' identities. This example encourages us to examine other frameworks that influence our practices in the Western culture in order to consider the intersections of other aspects of older adults' lives.

Box 14.1: The Canadian Model of Occupational Performance and Engagement (CMOP-E) and The Integrated Canadian Model of Occupational Performance and Engagement (ICMOP-E)

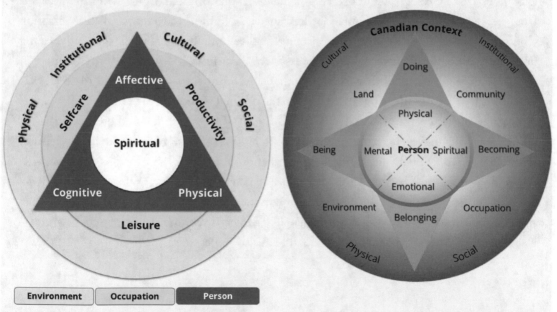

Left side image: Adapted from The Canadian Model of Occupation Performance (Polatajko et al., 2013)

Right side image: Adapted from Fijal and Beagan (2019, p. 226).

The Canadian Model of Occupational Performance and Engagement (CMOP-E) (Polatajko et al., 2013) was developed to promote client-centered practice and occupational therapy's view that occupation (meaningful activities) is the center of focus in terms of a client's overall health. This model includes three main components: person, occupation, and environment. The focal point of the CMOP-E is the interaction between the person, occupation, and environment and how this interaction can influence how a person is able to perform and engage in activities that are meaningful to them. For example, living on the top floor of an apartment with no elevator prevents an older adult with mobility issues from leaving home. Engagement, which is also embedded within the model, emphasizes the importance of clients being able to choose meaningful activities to focus on during therapy in an effort to provide the client with a sense of empowerment and improve therapy outcomes. By identifying what component(s) are preventing a client from being able to engage in certain activities, occupational therapists are able to design tailored programs that meet client needs.

Despite this model's aims of depicting individuals that are embedded within diverse cultural environments, its development takes on perspectives that are derived from the Western world (Hammell, 2011). As a result, it excludes populations that do not see the world through this lens, such as the Indigenous populations in Canada. In light of the Truth and Reconciliation Commission's recommendation of respecting and valuing Indigenous health and healing practices within the Canadian health care system (Truth and Reconciliation Commission of Canada, 2015), a model that includes Indigenous worldviews within the CMOP-E was developed by Fijal and Beagan (2019). Unlike the original version of the CMOP-E, the areas of occupation within the "integrated" CMOP-E (ICMOP-E) are focused on meaning (i.e., doing, being, belonging, becoming) as opposed to purpose (i.e., self care, productivity, leisure) (Hammell, 2014). Spirituality is the core foundation to all other dimensions when considering Indigenous health perspectives, therefore it is represented in the form of a medicine wheel at the center of the ICMOP-E. The importance of balance among physical, emotional, spiritual, and mental health among Indigenous peoples is represented in the circle, and the inseparability of person, community, and land are also represented.

PERSONA, SCENARIO, AND SOLUTION

Author: Adebusola Adekoya

Titi Torike

- **Age:** 72
- **Gender:** Woman, pronouns are she/her
- **Lives in:** Single family home with 82-year-old husband in Nigeria
- **Social circle:** Four adult children who are all married; two daughters live in Nigeria and two sons live in Canada. Active volunteer in her community and church but has been less active since the pandemic started.

Signature interests
Parent and grandparent, entertaining families and friends, women's rights activist.

" I want to make a positive difference in the lives of everyone I meet…Whatever I have, wherever I am, I can make it through anything in the One who makes me who I am (Philippians 4:13 MSG)"

Health
Good.

Technology
Not tech savvy. Connects with family and friends virtually through daughters' iPad and computer.

Persona. Titi is a 75-year-old African woman who lives with her 82-year-old husband, Tani, in a single-family home in Nigeria, West Africa. Titi is a retired teacher and Tani is a retired businessman. Titi has a medical history of hypertension and osteoarthritis. She had bilateral knee replacement about four years ago which helped her regain her mobility and physical activity level. Her hypertension is currently being treated with medications and lifestyle changes. She expresses the need to continue to make lifestyle changes by engaging in regular exercise and reducing sodium intake to enhance her quality of life. Tani's overall health is good. He has diabetes and mild cognitive impairment. His diabetes is currently being managed with diet but Titi notices that her husband is becoming more forgetful in the last six months.

Titi and Tani have four adult children. Their oldest son, who has Power of Attorney, recently immigrated to Edmonton, Canada with his family while the other son lives in Toronto. Both sons call Titi and her husband on their mobile phones (not smartphones) at least once a week and do video calls at least once a month. Her two daughters live in Nigeria with their families, about a 30-minute drive to Titi's place. Both daughters work full time and have young children between the ages of 2 and 10 and visit their parents at least once a week. Titi and her husband manage their daily activities with limited assistance from a carer who assists with house chores.

Scenario. Titi is a very active woman who loves going for a 30-minute walk every day in her neighbourhood. She values her autonomy and independence. She is an active member of her community, a women's rights activist and is dedicated to empowering women in her community. Titi is a Christian and her participation in religious and spiritual activities are important to her. Before the COVID-19 pandemic, she used to attend church services twice a week, community activities weekly, and host bi-weekly meetings with other women in the community to address the needs of women in her community and to socialize. Since the pandemic, some of the church services and community activities, including the meetings that she hosts, have been canceled. Most older adults in her community, including Titi and her husband, have yet to receive the COVID-19 vaccines, so Titi does not feel comfortable holding meetings in her home. She expresses feeling isolated and sometimes sad during the pandemic.

Recently, Titi has been reading about the benefits of technology and thinks it is time for her and the women in her community to embrace technology. She expresses the desire to connect more with her family, especially her grandchildren in Canada virtually without having to wait for her daughters to visit. She thought it would be a good idea to purchase her own iPad or tablet, but she cannot afford one, nor can she pay for the Internet. Also, she is worried about not being able to plug in the device to charge as power outages are frequent in her neighbourhood. She realizes that she needs to be trained on how to use the device. She also thinks about educating the women in her community about the need for technology and its benefits, including being able to host their bi-weekly meetings virtually, as needed. Titi is not sure how to address these issues and thinks that she might need her children's help but does not want to burden them.

Solution. To support Titi's need for technology and facilitate social connectedness with her family and friends, Titi needs access to technology and the Internet as well as a hands-on tutorial on how to use a mobile device. Although she expresses her concerns about being a burden to her children, Titi needs to share these concerns with her children, especially her need for financial assistance in purchasing a mobile device and paying for the Internet. If possible, Titi's daughters could provide a hands-on tutorial on how to use a mobile device to her and her husband.

To address her community's need for technology, Titi could partner with ICT organizations in Nigeria and local organizations for seniors and women to discuss the benefits of technology and potential solutions for making mobile devices and internet accessible to women and seniors in her community. Another approach is for the local or state government to subsidize the costs of mobile devices and internet subscriptions for women and seniors in Titi's community and have volunteers deliver hands-on tutorials on how to use the devices at no cost to them.

14.2   SOCIAL INEQUITIES AND THE DIGITAL DIVIDE

In Part III (Chapters 6–8), we examined the ways technologies can support self care, community contributions, enjoyment, and self-fulfillment. Despite the promise of these technologies, numerous examples of AgeTech solutions are not accessible to a large segment of the population due to a digital divide. The persona of Titi illustrates that there are many parts of the world where there is simply lack of access to the Internet and people must rely on help from family members. While the rates of older adults who use the Internet and have smartphones and other everyday digital technologies is steadily rising (AGE-WELL, 2021; Anderberg, 2020; Statistics Canada, 2019) not all older adults have access to these, nor assistive products and high tech innovations. Although disability can affect access to everyday technologies, income and education may be greater contributing factors to access and use (Kottorp et al., 2016). Similarly, older adults who have higher education levels and income are more likely to use the Internet than those with less education and income. In 2016, 89% of Canadian university-educated older adults regularly used the Internet compared to only 46% of Canadian older adults with a high school diploma (Davidson and Schimmele, 2019; Statistics Canada, 2019). Lack of access and use of the everyday digital technologies that can support older adults' performance of life activities can therefore negatively impact their independence. To take full advantage of the potential of technologies to facilitate independence, supports, policies, and programs must be in place for those without sufficient income to access such technologies. Otherwise, certain segments of the older adult population can be excluded from the use of these technologies, the activities that they support, as well as their benefits. The cost of the initial purchase of technologies and monthly service and subscription must be considered as well as maintenance and upgrade costs (Carnemolla, 2018); failing to account for both sets of costs creates a barrier to older adults' regular use of technologies.

Even if older adults are given access to hardware, software, subscriptions, and internet packages, such access alone will not result in the use and adoption of technologies to support their independence in self care. Although more older adults have undoubtedly improved their technology skills and confidence as a result of the COVID-19 pandemic (Morrow-Howell et al., 2020), training and support may still be required for many, especially when the technology is new to them. According to a 2015 Pew Research Center survey, only one-quarter of Internet users who were aged 65 and older stated that they were very confident using technology such as computers and smartphones to carry out the activities that they needed to do online (Anderson and Perrin, 2017). One-third indicated that they were only a little or not at all confident using technology for online activities. Similarly, almost three-quarters of older adults described needing help to set up new devices and be shown how to use them (Anderson and Perrin, 2017). In Titi's situation in Nigeria, organizational and governmental support is required to enable the communities to access technological solutions.

Successful AgeTech adoption depends on support and training for older adults, including training that caters specifically to persons with mild dementia or cognitive impairment (Patomella et al., 2018). Research also exists on the effectiveness of training programs to develop older adults' internet skills (Hunsaker and Hargittai, 2018). Libraries and senior centers can be a source of support for in-person and virtual support. The Waterloo Undergraduate Student Association's EnTECH club (http://www.entech.club/) is an example of a student-led initiative that provides such support to older adults, and there are many non-profit organizations that provide similar services such as Cyber-Seniors (https://cyberseniors.org/). In addition, a user-centered co-design approach can be used when developing instructions, resources, and tip sheets for older adults to ensure that these adequately meet the needs of the end users. Florida State University's Institute for Successful Longevity used such an approach to develop how-to guides for older adults to use Zoom (The Institute for Successful Longevity, 2020). It is strongly recommended that offline, such as in-person, as well as online training be made available (Xie et al., 2020). Moreover, for older adults from socially isolated ethnic groups, the expertise of community health workers could be leveraged to serve not only as cultural brokers but also as technology brokers (Xie et al., 2020). Such an approach could help reduce the gap between those older adults with and those without access to everyday technologies.

Figure 14.1: Older women learning to use the Internet on a PC at a computer training course. Shutterstock: Robert Kneschke.

PERSONA, SCENARIO, AND SOLUTION
Authors: Masako Miyazaki and Toshio Ohyanagi

Kazuko Hasegawa

- **Age:** 85
- **Gender:** Woman, pronouns are she/her
- **Lives in:** Matsudo, Japan
- **Social circle:** Widow, lives alone, has two adult disabled children who reside in nursing homes

Signature interests

Used to enjoy gardening and socializing, currently few interests due to limited physical capacity and loss of friends through deaths.

"I have lost all my friends and feel as though I am waiting to die."

Health

Frail, takes medications to manage cholesterol, blood pressure, and diabetes. Has memory decline, poor nutrition.

Technology

Uses PARO, a robotic baby seal that connects her to nurses and tele-support at a homecare agency.

Persona. Kazuko Hasegawa is 85 years old and lives alone in Matsudo, a commuter city east of Tokyo. She lives on the first floor of a walk-up four-story apartment. Although she is frail and gets confused easily, she is living independently with home care visits and tele-support. She has two children, but they are both living in nursing homes, one with dementia, and the other with Parkinson's. Kazuko asks her neighbour to get her food and supplies once a week.

Her friends have all passed or are too ill to communicate with her. But she enjoys talking to a robotic seal (PARO) which is connected to a home care monitoring center.

When Kazuko interacts with her robotic seal, the sensors communicate with the monitoring center. If she expresses pain, distress, an emotional outburst, or depression, information is transferred to the home care nursing station. If deemed necessary, a home care nurse is dispatched to visit Kazuko at home.

Scenario. Kazuko is living "independently" in her community with additional support. She is afraid of starting a fire by forgetting to turn off the stove. So, she goes to a corner store to buy ready-made Japanese food around 5:00 PM daily. As she can only carry a small amount of food, she purchases just the amount she needs and carries it home. Although there is also fast food nearby, she does not like Western food as it causes indigestion.

She often wonders if she has taken her medication for cholesterol, blood pressure, and diabetes. Occasionally, she forgets to eat breakfast and lunch. But daily, at 5:00 PM, she knows she must go out to get food from the corner store. Sometimes, she finds a large quantity of left-over, moldy food in her fridge. She cannot throw away food as it is wasteful. The home care nurse throws them out, otherwise, Kazuko would eat spoiled food and become ill. Once, when she got ill, she talked to the robotic seal which connected her to the call center. An ambulance was dispatched to take Kazuko to the hospital.

Kazuko wonders what will happen to her when she cannot take care of herself. From time to time, she asks herself if she has taken her pills, turned off the stove, and paid the bills. She has been told a few times that she called her pharmacist several times in one day. She has also been told that she forgot to pay a monthly utility bill. When she forgets to turn on the television, she loses track of the day and time.

Kazuko has no one to talk to except when she calls her children in the nursing home. But this is seldom as she finds it challenging to get through the switchboard in the nursing home.

Solution. There is little meaning and joy in Kazuko's life and she feels she is waiting for death to arrive. To address Kazuko's memory decline, the home care service programmed the robotic seal to remind Kazuko to take her pills and eat her meals.

On one visit, a home care nurse learned that Kazuko had enjoyed gardening and visiting friends in her younger years. As a trial and to compensate for her limited physical capacity, Kazuko was given a large tablet that allows her to visit and tour various gardens in Japan. But Kazuko found it challenging to navigate the Internet, so she quickly gave it up.

The home care service explored ways to help Kazuko. As Kazuko was proficient at interacting with the robotic seal, home care service programmed the seal to receive instructions from Kazuko about places she would like to see or visit. This information was then used by the call centre to remotely set up Kazuko's tablet, so she could enjoy using it to travel virtually to gardens with minimal effort. Over time, Kazuko was able to visit famous tree and flower gardens, trendy coffee shops, and virtual walking tours of famous sites around the world.

She continues to be firmly bonded with her robotic seal which gives her joy and facilitates her ability to live at home independently.

14.3 EVOLUTION OF TECHNOLOGIES

In Part II (Chapters 3–5), we examined how technologies are tools that can support an older adult's sense of self and identity through digital storytelling and assisting with daily tasks when one's memory is declining. Emerging technologies can give older adults, even those with declining cognitive functions, autonomy over their personal data, preferably if they have appointed trusted guardians a priori. With technology, relationships between older people and their caregivers can be facilitated and allow a degree of independence and peace of mind, as in the case of monitoring. Technology can also facilitate social support and connection when people are physically separated.

As technologies become ubiquitous, norms and expectations change. For example, smartphones allow users to easily contact others without the need to remember phone numbers. It is no longer considered a sign of cognitive decline if users are "dependent" on their devices for reaching someone. We also rely on smart mobile devices to remind us of appointments, and to help us navigate when we drive. The skills to remember phone numbers or to read a map are no longer expected of average citizens.

Risks of using technology include privacy and security breaches that can make an older adult vulnerable to crime. Older adults who are continuously monitored may have their privacy infringed upon. Having one's home outfitted with various sensors can mean that information about an individual's day-to-day activities and routines is freely shared with remote caregivers, staff of assisted living or continuing care facilities, and health professionals. Older adults may lose the right to control access to their personal data. Privacy was identified as one of the most important factors that affect the adoption of smart home and remote monitoring technologies (Berridge and Wetle, 2020; Chung et al., 2016). However, evidence suggests that older adults may have positive attitudes toward in-home sensors when the technologies are unobtrusive (i.e., not physically or psychologically prominent) (Hensel et al., 2006), there is a perceived need for them, and they can support older adults to age in place (Townsend et al., 2011). Regardless of the benefits of these technologies, respect for older adults' autonomy requires that their decisions to use a technology, or be subject to a caregiver's use of technology while in their care, are negotiated ahead of time. There is a balance between this form of relational autonomy and risks.

As personal data becomes decentralized, as in the case of self-sovereign identity, older adults can have more control over their own data, who can use it and how they want to share it. It remains a challenge to determine the nature of the interface(s) for emerging technologies such as self-sovereign identity. Another challenge is to understand emerging technologies, such as artificial intelligence (AI), that utilize the data created using technology itself.

14.3.1 ARTIFICIAL INTELLIGENCE

As AI features more prominently in the services and interventions offered to older people, an emerging issue relates to the validity of the recommendations or "decisions" made by a technology that uses AI. AI is based on data, which is the collection of samples of a phenomenon. The sources of data can be based in nature, or created by humans (Burkov, 2019). Examples of data produced by humans are databases that contain historical information about processes or established practices. This historical information is stored somewhere and can be used to classify things and make predictions of, or decisions about, the future.

Humans have conscious and unconscious biases that advantage or disadvantage certain groups of people. Negative biases further disadvantage marginalized populations based on their race, age, gender, economic status, migratory status, and nationality, among many other attributes. Historically, data is permeated with human bias, and when this is not recognized and addressed, AI inherits these biases, which results in exclusion of certain social groups from services that they require.

There are many examples of biases in AI. The purpose of this chapter is not to present an exhaustive demonstration of such examples but to raise the issue that since AI uses data to make classifications or predictions, socially constructed data is inherently biased. One recent example is related to biases found in AI algorithms developed for facial recognition. Facial recognition uses Facial Processing Technology, which includes a variety of processes from face detection (computer distinguishes a face from a background), facial analysis (computer identifies physical or demographic traits in an individual's face), and face verification or identification (computer recognizes the face of a specific individual). Facial Processing Technology can be used for innocuous purposes such as smile detection; however, it can also be used as a biometric system for tracking individuals. If systems are intentionally or unintentionally fed with biased data, they can discriminate against marginalized groups (Raji et al., 2020). Raji and colleagues analyzed the performance of three commercially available facial recognition technologies (Microsoft, Amazon, and Clarifai) and found that for the intersectionality of skin type (dark/light) and gender (women/men), all these technologies showed lower classification accuracy for darker women, while the highest accuracy was for the lighter men. This misclassification is dangerous as it has led to the misidentification of suspects of crime.

The issue of biased AI has been brought to attention by several prominent researchers well known in academia and in the media such as Meredith Broussard (2018), who questions whether or not we can be optimistic about the power of AI. Cathy O'Neil (2016) also raises the issue that many AI algorithms being used today are not transparent and reinforce discrimination; for example, if a student who lives in a poor neighborhood is labelled by an algorithm as being "too risky" to enter in a college because of the student's postal code, this person misses education opportunities to get out of poverty.

We need to be aware of the social problems that come with technology, such as AI, as these technologies are used to make decisions about our lives and future in aspects such as finances, education, and health (Broussard, 2018). Biases in AI lead to inequities in access to education, jobs, health services, and credit, among other services. Every effort is needed to remove biases in algorithms used in AI. One way is to diversify developers of these algorithms so that the data used represents the developers and the populations intended to be served by innovation. Finally, developers of AI algorithms need to transparently communicate the limit of the dataset used and its appropriate context of use (Raji et al., 2020).

With the cautions of AI aside, AI holds much promise for many applications that can support autonomy and independence for older adults as demonstrated in the above persona, scenario and solution featuring Kazuko Hasegawa, a frail elderly woman who lives in Japan.

14.4   THE IMPORTANCE OF POLICIES TO CREATE AND SUSTAIN CHANGES

Much of the content of this book has been influenced by our collective experiences during the global pandemic. Technologies have received heightened attention since the beginning of the global pandemic in early 2020. Requirements for quarantine and physical distancing has made it necessary for older adults, their social circle, and service providers to resort to technologies for communication, virtual contact, and access to amenities. For some experts, the necessity of using technology for receiving and providing services simply became possible when decision makers "got out of the way." For example, technology was not an issue for telehealth; the issue was billing for services that were delivered virtually, by telephone or video conferencing, versus in person. When professionals could bill services delivered virtually, we saw that the number or proportion of users increased dramatically (DeSilva et al., 2021; Lin et al., 2018; Sisk et al., 2020). An emerging issue is whether gains we have made with technologies during the pandemic will become "legacy" practices. Will policies for practice changes that are for the better, be maintained after the pandemic is over? According to experts, such as Louise Schaper, CEO of Australasian Institute of Digital Health, long-term implementation of improved practices require will power and collaboration of policy makers and all stakeholders (Schaper, 2021).

To address the digital divide exacerbated by the pandemic, policies are needed to make telecommunication costs affordable. Telecom prices in Canada are the highest in the world (Rewheel Research, 2020; Wall Communications Inc., 2019). This high cost makes it challenging for health innovations to be accessible and affordable to everyday consumers. It also makes it prohibitive for businesses to commercialize their products in Canada.

Policies are also needed to create an environment for flexibility and accessible care, some of which was demonstrated during the pandemic. Currently, Amazon Care in the United States delivers convenient care to the home of citizens (https://amazon.care/). In Canada, for a fee or subscription, Maple (https://www.getmaple.ca/) provides online visits with physicians and prescriptions. Currently, during the pandemic, mental health consultations through Maple are covered by one provincial health plan. Again, the technologies are available, the challenge is the creation, implementation and maintaining of policies that bring accessible and convenient services to citizens in the comfort and safety of their homes.

14.5   FUTURE DIRECTIONS

From the challenges described in this chapter, we propose the following areas of focus to further drive the field of AgeTech as it pertains to autonomy and independence.

- Reconceptualize the terms "autonomy" and "independence."

Humans are social beings, and each individual has a sense of self with decision-making capacity. In an ageist society, aging is associated with dependency. If we accept that no one is completely independent and that people are "interdependent" on their human and non-human environments, we can accept that with aging, to be independent refers to having a choice in the type of help people receive. People make these choices based on their relationships with their social circles and their environment, hence, "relational autonomy" would provide better understanding of an individual's situation.

• Address social inequities in the design, implementation, and adoption of innovations targeted at older adults. This could be supported by the use of co-creation and co-design approaches involving older adults as partners in the process of developing new AgeTech.

The key is to ensure that target end users include under-represented segments of our populations. Characteristics of under-represented groups include considerations for gender, socio-economic status, language, and other cultural factors. If we do this well, we would address issues of access, affordability and other factors that contribute to the digital divide.

• Anticipate changes in expectations and norms that come with technology evolutions.

As innovations continue to grow at exponential rates, there are changes to the ways we live, work, play, socialize, deliver and obtain services, include health care. The adage, "the only thing that doesn't change is change itself" applies here. This means that we can expect changes to "gold standards," "best practices," and assumptions of what older adults can and cannot do based on technology solutions that emerge in the future.

• Include policies and practices in the innovation ecosystem.

Just as technologies evolve, so should our policies and practices. Changes in policies and business practices are also recognized as innovation. For example, the standard two-year telecommunication contract is not appropriate for a person living with progressive cognitive decline using a wearable GPS device when the device may only be used for a few months. The development of business incentives to foster the engagement of seniors in innovation processes (seniorpreneurs) would be innovative. As world populations continue to age, more seniors will be seeking opportunities to remain engaged within their communities and will be seeking jobs that transcend beyond their retirement years. We can include their experience and enthusiasm to foster entrepreneurial ventures.

Glossary

Activity: The ICF defines activity as "the execution of a task or action by an individual" (World Health Organization, 2002, p. 10). Examples of activities are learning and applying knowledge, grasping objects, starting a conversation, or toileting.

Activity limitations: According to the ICF, activity limitations occur when a person experiences difficulties in performing activities. Examples of activity limitations are difficulties in carrying out daily routines, sustaining a conversation, or handling objects.

Advance directives or advanced care planning: A way for persons to ensure that one's wishes and values are respected especially related to health decisions. These are "an effective means of extending an individual's autonomy from their current self, as an individual who has decisional capacity, onto their future self, who lacks it" (Walsh, 2020, p. 54).

AgeTech: The use of advanced technologies such as information and communication technologies (ICTs), robotics, mobile technologies, artificial intelligence (AI), ambient systems, and pervasive computing to drive technology-based innovation to benefit older adults (Pruchno, 2019).

Autonomy: Originated in Greek and referred to self-rule, self-determination, free will, and self-sovereign (Dryden, 2014; Dworkin, 2015; Oxford Learner's Dictionary, 2021). Autonomy is a term used broadly to refer to liberty, self-rule or sovereignty, and freedom of the will (Dworkin, 2015). It can also equate to "dignity, integrity, individuality, independence, responsibility and self-knowledge" (Dworkin, 2015, p. 8). Autonomy is considered an individual right, and respect for autonomy is one of the ethical principles in health care (see Box 1.1). Autonomy also refers to an ability to make moral decisions guided by one's moral codes.

Basic (or personal) activities of daily living (ADL): These include care and hygiene tasks such as toileting and management of continence, dressing, grooming, feeding, bathing/showering, and basic mobility in bed and in one's immediate surroundings (Jacobs and Simon, 2015; Matuska and Christiansen, 2011; Polatajko et al., 2007; Whitehead et al., 2013).

Body functions: Physiological functions of body systems, including psychological functions, for example, thinking, remembering, pain, and functions of the cardiovascular system (World Health Organization, 2002).

Body structures: Anatomical parts of the body, for example, the brain, the eyes, structures related to the movement such as bones or muscles, or the skin (World Health Organization, 2002).

Capacity: The term capacity has been understood in different ways at different times and places. There are multiple terms, such as "mental capacity," "competence," "decision-making ability" (and their opposites, including "mental incapacity" and "incompetence"), which are sometimes used as synonyms or near synonyms of "capacity," and are used to make important conceptual distinctions.

Co-decision making: Also known as supported or assisted decision making, is another model of guardianship within Canada where a co decision-maker assists the adult in making decisions, rather than making decisions for them.

Community mobility: Moving oneself within the community using public or private transportation (American Occupational Therapy Association, 2002, p. 620).

Digital divide: Refers to discrepancies in access to information communication technologies, or the internet, between segments of our populations (Fang et al., 2018; Hilbert, 2011).

Disability: Refers to a condition that is physical, sensory, cognitive, or psychological that affects a person's level of function in daily activities. Disabilities can be temporary, chronic, reversible, or progressive. Aging is associated with disabilities.

Enhanced activities of daily living: Activities are done for the purposes of fun, pleasure, relaxation, learning or growth, play, leisure, recreation, and entertainment (Rogers et al., 1998, 2020).

Ethics of care: Emphasizes the importance of relationships between the caregiver and care recipient, provides a commitment to effective communication, and one that appreciates the uniqueness of individuals in unique situations (Hughes and Baldwin, 2006).

Everyday digital technologies: High tech that we use to engage in everyday activities, and to connect with people and the environment. These technologies include products such as smart devices (e.g., smartphones, smartwatches, smart homes), mobile devices (cell phones, tablets, computers), and digital communication platforms and software and algorithms. They also include services and policies regarding the cost, literacy, and protection of personal information associated with the technology's access and use.

Exergames: Also known as virtual reality training and exertion games, exergames combine video game technology with sensors worn on the body or placed near the user to complete on-screen activities ranging from simulated sports (e.g., bowling, skiing) to adventure tasks (e.g., battling dragons) (Donath et al., 2016; Kappen et al., 2019; Zheng et al., 2019).

Goldilocks Principle on Dementia and Wayfinding: This principle highlights the need for the perception of risk of dementia-related wandering being "just right" among caregivers, persons living with dementia and health professionals.

Guardianship: Grants legal authority to a care partner to make decisions on behalf of an older adult when one no longer has the capacity to do so. It is an option available when a person no longer has a power of attorney or an advanced directive already in place. Guardians can make personal, non-financial decisions about health care, where to live, who to associate with, participation in social activities, employment, and legal proceedings. Guardianship cannot make decisions about finances, and organ donations. A guardianship over an older adult lasts until the older adult regains the ability to care for oneself, or until the older adult passes away.

Harm: Any physical injury or damage to the health of people, or damage to property or the environment (International Organization for Standardization, 2007, p. 4).

Harm reduction: Pertains to risk and autonomy among older adults and persons living with dementia. Harm reduction can include any strategy that is designed to reduce harm or the probability of something negative happening, without requiring the complete cessation of the risky behavior.

Hazard: A potential source of harm.

Hazardous situations: Those "circumstances in which people, property, or the environment are exposed to one or more hazard(s)" (International Organization for Standardization, 2007, p. 4).

Impairments: Significant problems in body structure or function.

Implementation research: "The scientific study of methods to promote the systematic uptake of research findings and other evidence-based practices into routine practice, and, hence, to improve the quality and effectiveness of health services" (Eccles and Mittman, 2006, p. 1).

Independence: The degree to which a person can perform a task or activity or perform a role.

Individual autonomy: Embodies the mainstream understanding of autonomy where one has the capacity and freedom to make your own choices, as long as one understands the choice being made (Department for Constitutional Affairs, 2007).

Instrumental activities of daily living (IADLs): Activities that typically involve greater cognitive demands than basic activities of daily living and thus are considered more advanced tasks (Czarnuch and Mihailidis, 2011).

The International Classification of Functioning, Disability and Health (ICF): A classification of health and health-related domains, including the biological, individual, and social aspects of health. It proposes a biopsychosocial model to understand disability. The ICF describes optimal functioning and disability as outcomes of the interaction between the health condition and contextual factors. The health condition includes the presence and severity of diseases, disorders, and injuries, while contextual factors include personal and environmental factors. Based on the biopsychosocial model, the ICF identifies three levels of human functioning classified in functioning at the level of the body or body part, the whole person, and the whole person in a social context.

Machine learning: As a part of artificial intelligence, machine learning involves computers that learn, predict, and make decisions similar to humans (Alpaydin, 2020).

Memory prosthesis: Devices that facilitate interactions with family and friends, helping to maintain connection and enjoyment in interacting with one another despite memory and communication challenges (Gelonch et al., 2019).

Participation: Defined as "involvement in a life situation" (World Health Organization, 2002, p. 10), in other words, participation in performing activities associated with social life roles.

Participation restrictions: When a person experiences limited or lack of access to opportunities, for example, the social stigma of HIV can contribute to unemployment, or an inability to use public transportation, or limited education.

Perceived risk: Probability of harm or undesirable experience that is created by society and is mediated by social and cultural processes (Lupton, 2005).

Power of attorney: Grants legal authority to an individual to make decisions on behalf of another. In Canada, a power of attorney is a legal document that allows a person's designated attorney to manage their finances and property on their behalf when they lack the capacity to manage their own affairs (Government of Canada, 2016).

Privacy: As described by Westin (1967), the freedom to disclose (or not disclose) personal information without incurring unwanted social control by others. Privacy has historically been conceptualized as a static term as only involving intimate actions such as toileting and bathing tasks; however, it also enables a person to engage in the creation of "self" and has the capability to fulfil a variety of psychological needs, such as intimacy, anonymity, and solitude (McNeill et al., 2017).

Relational autonomy: An alternative approach to autonomy which requires that attention is paid to the ways that people exist within relations of social support and community (Nedelsky, 1989).

Risk: The probability of occurrence of harm and the consequences of that harm, in other words, how severe the harm might be (International Organization for Standardization, 2007, p. 4).

Risk management strategies: A "systematic application of management policies, procedures and practices to the tasks of analyzing, evaluation, controlling and monitoring risk" (International Organization for Standardization, 2007, p. 4).

Sequence of events: A hazard that cannot result in harm until such time as a sequence of events lead to a hazardous situation.

Smart assistant technologies: Also known as smart assistants (Seymour, 2018), virtual assistants (Bello et al., 2018), digital assistants (Trajkova and Martin-Hammond, 2020), voice-controlled personal assistants (O'Brien et al., 2020), intelligent voice assistants (Polyakov et al., 2018), and voice-enabled technologies (Jadczyk et al., 2019), smart assistant technologies use artificial intelligence algorithms to perform different tasks, respond to commands and questions, provide users with information, and control other connected electronics.

Social robots: Also known as socially interactive robots, social robots are high tech solutions capable of having meaningful interactions with people (Sandry, 2015; Sarrica et al., 2020).

Technology: Created by humans to enhance the performance of tasks or functions (Grübler, 2003). The concept of technology includes a combination of "hardware," which refers to tangible human-made things (e.g., buildings, equipment) and "software," which is the organized arrangement of human knowledge, experience, and skills (Li-Hua, 2012). Simply put, technology includes things, actions, processes, methods, and systems made by humans (Kline, 1985). The term "technology" is used to describe a range of products or services, such as high technology, or "high tech," referring to advanced technology, frontier technology or cutting-edge technology, which are the newest or most modern machines, equipment, or methods available.

Technology adoption: Refers to the time it takes for a group of individuals (the adopters) to incorporate new technology in their lives consistently. The technology adoption by each individual of a social group follows a process five steps: (1) the individual becomes aware of a new technology's existence (awareness); (2) the individual finds out enough information about the technology's features (persuasion); (3) the individual makes the decision whether to adopt or reject the technology (decision making process); (4) the individual acts accordingly on his or her decision (implementation); and (5) the individual engages in an ongoing assessment process on whether to continue or discontinue with the technology adoption decision (confirmation) (Rogers, 1995) cited by Straub (2009).

Trusteeship: A situation in which a person's money or property is managed by another person.

Usability: A concept associated with the actual use of technology, based on the user experience when interacting with technology. Usability is a multidimensional concept with five attributes, namely: learnability (easy to learn the technology), efficiency (level of productivity achieved with minimum required resources), memorability (easy to remember how to use the technology), errors (free of error or low error rate), and satisfaction (pleasant to use) (Nielsen, 1993).

References

AARP and National Alliance for Caregiving. (2020). *Caregiving in the United States 2020*. DOI: 10.26419/ppi.00103.001. 11

Active Aging in Manitoba. (2020). Staying active during Covid-19. https://activeagingmb.ca/resources/staying-active-during-covid-19/. 103

Adams, K. B., Leibbrandt, S., and Moon, H. (2011). A critical review of the literature on social and leisure activity and wellbeing in later life. *Ageing and Society*, 31(4), 683–712. DOI: 10.1017/S0144686X10001091. 110

Adekoya, A. A. and Guse, L. (2019). Wandering behavior from the perspectives of older adults with mild to moderate dementia in long-term care. *Research in Gerontological Nursing*, 12(5), 239–247. DOI: 10.3928/19404921-20190522-01. 70

Adekpedjou, R., Stacey, D., Brière, N., Freitas, A., Garvelink, M. M., Turcotte, S., Menear, M., Bourassa, H., Fraser, K., Durand, P. J., Dumont, S., Roy, L., and Légaré, F. (2018). "Please listen to me": A cross-sectional study of experiences of seniors and their caregivers making housing decisions. *PLOS ONE*, 13(8), e0202975. DOI: 10.1371/journal.pone.0202975. 31

Adler, G. (2010). Driving decision-making in older adults with dementia. *Dementia*, 9(1), 45–60. DOI: 10.1177/1471301209350289. 71, 72

Adler, R. P. and Goggin, J. (2005). What do we mean by "civic engagement"? *Journal of Transformative Education*, 3(3), 236–253. DOI:10.1177/1541344605276792. 109

AGE-WELL. (n.d.). Core research program awards 2020–2023. https://agewell-nce.ca/wp-content/uploads/2020/07/AGE-WELL-Core-Research-Program-and-Platform-Projects-2020-2023-Project-Summaries-1.pdf. 109

AGE-WELL. (2021). COVID-19 has significantly increased the use of many technologies among older Canadians: Poll. https://agewell-nce.ca/archives/10884. 14, 97, 99, 103, 165

AGE Platform Europe. (2021). What is it to be a grandparent during COVID-19? https://www.age-platform.eu/policy-work/news/what-it-be-grandparent-during-covid-19. 114

Age UK. (2021). Seniors dating in later life. https://www.ageuk.org.uk/information-advice/health-wellbeing/relationships-family/dating-in-later-life/. 126

Aging 2.0. (2021). Day 18 - Tokyo - Mimamoriai. https://www.aging2.com/blog/day-18-to-kyo-mimamoriai/. 139

Agli, O., Bailly, N., and Ferrand, C. (2015). Spirituality and religion in older adults with dementia: A systematic review. *International Psychogeriatrics*, 27, 715–725. DOI: 10.1017/S1041610214001665. 129

Ahluwalia, S. C., Gill, T. M., Baker, D. I., and Fried, T. R. (2010). Perspectives of older persons on bathing and bathing disability: A qualitative study. *Journal of the American Geriatrics Society*, 58(3), 450–456. DOI: 10.1111/j.1532-5415.2010.02722.x. 86

Aleccia, J. (2020). Sheltered at home, families broach end-of-life planning. *Kaiser Health News*. https://khn.org/news/coronavirus-medical-directives-end-of-life-planning/. 56

Alharbi, M., Straiton, N., Smith, S., Neubeck, L., and Gallagher, R. (2019). Data management and wearables in older adults: A systematic review. *Maturitas*, 124, 100–110. DOI: 10.1016/j.maturitas.2019.03.012. 100

Allemann, H. and Poli, A. (2020). Designing and evaluating information and communication technology-based interventions? Be aware of the needs of older people. *European Journal of Cardiovascular Nursing*, 19(5), 370–372. DOI: 10.1177/1474515119897398. 147

Alpaydin, E. (2020). *Introduction to Machine Learning*, 4th ed. The MIT Press. 131, 176

Alsaqer, M. and Hilton, B. (2015). Indirect wayfinding navigation system for the elderly. *Americas Conference on Information Systems*, 1–13. https://core.ac.uk/download/pdf/301365857.pdf. 135

Alzheimer's Association. (n.d.). Respect for autonomy. https://www.alz.org/documents/national/autonomyei.pdf. 44, 51, 136

Alzheimer Scotland. (2018). NHS Grampian encourage people to download Purple Alert. https://www.alzscot.org/news/nhs-grampian-encourage-people-to-download-purple-alert. 137, 138

Alzheimer Scotland. (2019). Missing person app for people with dementia hits 10,000 downloads. https://www.alzscot.org/news/missing-person-app-for-people-with-dementia-hits-10000-downloads. 137, 138

Alzheimer Scotland. (2021a). Purple Alert FAQ. https://www.alzscot.org/purplealert/FAQ.

Alzheimer Scotland. (2021b). Purple Alert privacy notice. https://www.alzscot.org/purplealert/privacy. 137

Alzheimer Society of Canada. (2021). Managing the changes in your abilities. https://alzheimer.ca/en/help-support/im-living-dementia/managing-changes-your-abilities. 44

Alzheimer Society of London & Middlesex. (2021). Reframing dementia as a disAbility. https://alzheimerlondon.ca/bm-1-19-21/. 10

Alzheimer Society of Toronto. (2021). Our connections matter. https://on.alz.to/site/SPage-Server/?pagename=AST_Connections_that_Matter. 103

American Occupational Therapy Association. (2002). Occupational therapy practice framework: Domain and process. *American Journal of Occupational Therapy*, 56, 609–639. DOI: 10.5014/ajot.56.6.609. 135, 174

American Red Cross. (2021). Be a digital advocate. https://www.redcross.org/volunteer/volunteer-opportunities/be-a-digital-advocate.html. 112

Anderberg, P. (2020). Gerontechnology, digitalization, and the silver economy. *XRDS: Crossroads, The ACM Magazine for Students*, 26(3), 46–49. DOI: 10.1145/3383388. 165

Anderson, M. and Perrin, A. (2017). Tech adoption climbs among older americans. In *Pew Research Center: Internet, Science & Technology*. https://www.pewresearch.org/internet/2017/05/17/tech-adoption-climbs-among-older-adults/. 126, 165

Anderson, S. and Loeser, R. F. (2010). Why is osteoarthritis an age-related disease? *Best Practice and Research: Clinical Rheumatology*, 24(1), 15–26. DOI: 10.1016/j.berh.2009.08.006. 85

Appel, L., Appel, E., Bogler, O., Wiseman, M., Cohen, L., Ein, N., Abrams, H. B., and Campos, J. L. (2020). Older adults with cognitive and/or physical impairments can benefit from immersive virtual reality experiences: A feasibility study. *Frontiers in Medicine*, 6, 329. DOI: 10.3389/FMED.2019.00329. 104

Arriagada, P. (2018). A day in the life: How do older Canadians spend their time? https://www150.statcan.gc.ca/n1/pub/75-006-x/2018001/article/54947-eng.htm. 119, 125

Arriagada, P. (2020). The experiences and needs of older caregivers in Canada. https://www150.statcan.gc.ca/n1/en/pub/75-006-x/2020001/article/00007-eng.pdf?st=yfE-XL3j. 9, 11, 114

Arthanat, S. (2019). Promoting information communication technology adoption and acceptance for aging-in-place: A randomized controlled trial. *Journal of Applied Gerontology*, 1–10 DOI:. 10.1177/0733464819891045.

Aston, J., Vipond, O., Virgin, K., and Youssouf, O. (2020). Retail e-commerce and COVID-19: How online shopping opened doors while many were closing. *Statistics Canada*. https://www150.statcan.gc.ca/n1/pub/45-28-0001/2020001/article/00064-eng.htm. 90

Aud, M. A. (2004). Dangerous wandering: Elopements of older adults with dementia from long-term care facilities. *American Journal of Alzheimer's Disease and Other Dementias*, 19(6), 361–368. DOI: 10.1177/153331750401900602. 72

Ault, L., Goubran, R., Wallace, B., Lowden, H., and Knoefel, F. (2020). Smart home technology solution for night-time wandering in persons with dementia. *Journal of Rehabilitation and Assistive Technologies Engineering*, 7. DOI: 10.1177/2055668320938591. 94, 95

AUTM. (2011). AUTM. https://autm.net/source/STATT/. 149

Azad Khaneghah, P. (2020). Alberta rating index for apps (ARIA): An index to rate the quality of mobile health applications. Doctoral dissertation, University of Alberta. DOI: 10.7939/R3-QAGM-6984. 129

Bäccman, C., Bergkvist, L., and Kristensson, P. (2020). Elderly and care personnel's user experiences of a robotic shower. *Journal of Enabling Technologies*, 14(1), 1–13. DOI: 10.1108/JET-07-2019-0033. 86

Baric, V., Andreassen, M., Öhman, A., and Hemmingsson, H. (2019). Using an interactive digital calendar with mobile phone reminders by senior people - A focus group study. *BMC Geriatrics*, 19(1), 116. DOI: 10.1186/s12877-019-1128-9. 113

Bartlett, C., Marshall, M., and Marshall, A. (2012). Two-eyed seeing and other lessons learned within a co-learning journey of bringing together indigenous and mainstream knowledges and ways of knowing. *Journal of Environmental Studies and Sciences*, 2(4), 331–340. DOI: 10.1007/s13412-012-0086-8. 5

Battiste, M. (2005). Indigenous knowledge: Foundations for first nations. *WINHEC: International Journal of Indigenous Education Scholarship*, 1, 1–17. https://journals.uvic.ca/index.php/winhec/article/view/19251. 12, 13

Beer, C., Lowry, R., Horner, B., Almeida, O. P., Scherer, S., Lautenschlager, N. T., Bretland, N., Flett, P., Schaper, F., and Flicker, L. (2011). Development and evaluation of an educational intervention for general practitioners and staff caring for people with dementia living in residential facilities. *International Psychogeriatrics*, 23(2), 221–229. DOI: 10.1017/S104161021000195X. 69

Beier, M. E., Torres, W. J., and Beal, D. J. (2020). Workplace aging and jobs in the twenty-first century. In Czaja, S. J., Sharit, J., and James, J. B. Eds., *Current and Emerging Trends in Aging and Work*. Springer International Publishing, pp. 13–32. DOI: 10.1007/978-3-030-24135-3_2. 108

Bello, L. Lo, Iannizzotto, G., Nucita, A., and Grasso, G. M. (2018). A vision and speech enabled, customizable, virtual assistant for smart environments. *2018 11th International Conference on Human System Interaction* (HSI), 50–56. DOI: 10.1109/HSI.2018.8431232. 177

Bennett, B. (2019). Technology, ageing and human rights: Challenges for an ageing world. *International Journal of Law and Psychiatry*, 66(September 2018), 101449. DOI: 10.1016/j.ijlp.2019.101449. 73

Beringer, R. M., Gutman, G., and Vries, B. de. (2021). Exploring and promoting advance care planning among lesbian, gay, bisexual, transgender, and queer (LGBTQ) older adults living in non-metropolitan British Columbia. *Journal of Gay and Lesbian Social Services*, 33(4), 407–426. DOI: 10.1080/10538720.2021.1885552. 51, 59

Bernier, L. and Regis, C. (2019). Improving advance medical directives: Lessons from Quebec. https://irpp.org/research-studies/improving-advance-medical-directives-lessons-from-quebec/. 57

Berridge, C. and Wetle, T. F. (2020). Why older adults and their children disagree about in-home surveillance technology, sensors, and tracking. *The Gerontologist*, 60(5), 926–934. DOI: 10.1093/geront/gnz068. 77, 169

Bewernitz, M., Mann, W. C., Dasler, P., Otr, M. A., and Belchior, P. (2009). Feasibility of machine-based prompting to assist persons with dementia. *Assistive Technology*, 21(4), 196–207. DOI: 10.1080/10400430903246050. 87

Bhatt, J., Walton, H., Stoner, C. R., Scior, K., and Charlesworth, G. (2020). The nature of decision-making in people living with dementia: A systematic review. *Aging and Mental Health*, 24(3), 363–373. DOI: 10.1080/13607863.2018.1544212. 48, 71

Biagi, E. S. L. (2021). Smart home report 2021. https://www.statista.com/study/42112/smart-home-report/.

Blocker, K. A., Wright, T. J., and Boot, W. R. (2014). Gaming preferences of aging generations. *Gerontechnology: International Journal on the Fundamental Aspects of Technology to Serve the Ageing Society*, 12(3), 174. DOI: 10.4017/gt.2014.12.3.008.00. 130

Boddy, J., Chenoweth, L., McLennan, V., and Daly, M. (2013). It's just too hard! Australian health care practitioner perspectives on barriers to advance care planning. *Australian Journal of Primary Health*, 19(1), 38–45. DOI: 10.1071/PY11070. 51

Boman, I. L., Bartfai, A., Borell, L., Tham, K., and Hemmingsson, H. (2010). Support in everyday activities with a home-based electronic memory aid for persons with memory impairments. *Disability and Rehabilitation: Assistive Technology*, 5(5), 339–350. DOI: 10.3109/17483100903131777. 113

Bonifazi, W. L. (2000). Out for a walk. Can wandering be redirected into positive activity? Here's how to quell the wanderlust. *Contemporary Longterm Care*, 23(9), 40–46. http://www.ncbi.nlm.nih.gov/pubmed/11187298. 68

Boot, W. R. (2020). Technology to support leisure activities. *Gerontechnology*, 19(0), 1–1. DOI: 10.4017/gt.2020.19.s.69681.5. 130

Boot, W. R., Andringa, R., Harrell, E. R., Dieciuc, M. A., and Roque, N. A. (2020). Older adults and video gaming for leisure: Lessons from the Center for Research and Education on Aging and Technology Enhancement (CREATE). *Gerontechnology*, 19(2), 138–146. DOI: 10.4017/gt.2020.19.2.006.00. 120, 130

Borg, J., Larsson, S., and Östergren, P. (2011). The right to assistive technology: For whom, for what, and by whom? *Disability & Society*, 26(2), 151–167. DOI: h10.1080/09687599.2011.543862. 25, 26

Bostic, S. M. and McClain, A. C. (2017). Older adults' cooking trajectories: Shifting skills and strategies. *British Food Journal*, 119(5), 1102–1115. DOI: 10.1108/BFJ-09-2016-0436. 88

Botham, R. and Graves, A. (2009). Third age entrepreneurs: Innovative business start ups in mid-life and beyond – understanding the drivers and removing the barriers. In *Interim report to NESTA*. 152

Bouchard, B., Bouchard, K., and Bouzouane, A. (2020). A smart cooking device for assisting cognitively impaired users. *Journal of Reliable Intelligent Environments*, 6(2), 107–125. DOI: 10.1007/s40860-020-00104-3. 90

Bowes, A. and McColgan, G. (2013). Telecare for older people: Promoting independence, participation, and identity. *Research on Aging*, 35(1), 32–49. DOI: 10.1177/0164027511427546. 39

Brickwood, K.-J., Williams, A. D., Watson, G., and O'Brien, J. (2020). Older adults' experiences of using a wearable activity tracker with health professional feedback over a 12-month randomised controlled trial. *Digital Health*, 6, 1–13. DOI: 10.1177/2055207620921678. 101

Brittain, K., Corner, L., Robinson, L., and Bond, J. (2010). Ageing in place and technologies of place: The lived experience of people with dementia in changing social, physical and technological environments. *Sociology of Health & Illness*, 32(2), 272–287. DOI: 10.1111/j.1467-9566.2009.01203.x. 68

Broussard, M. (2018). Artificial Unintelligence: How Computers Misunderstand the World. Cambridge, MA: The MIT Press, p. 248. DOI: 10.7551/mitpress/11022.001.0001. 170

Brown, T., Findlay, M., von Dincklage, J., Davidson, W., Hill, J., Isenring, E., Talwar, B., Bell, K., Kiss, N., Kurmis, R., Loeliger, J., Sandison, A., Taylor, K., and Bauer, J. (2013). Using a wiki platform to promote guidelines internationally and maintain their currency: Evidence-based guidelines for the nutritional management of adult patients with head and neck cancer. *Journal of Human Nutrition and Dietetics*, 26(2), 182–190. DOI: 10.1111/jhn.12036. 65

Buck, H., Pinter, A., Poole, E., Boehmer, J., Foy, A., Black, S., and Lloyd, T. (2017). Evaluating the older adult experience of a web-based, tablet-delivered heart failure self-care program using gerontechnology principles. *Geriatric Nursing*, 38(6), 537–541. DOI: 10.1016/j.gerinurse.2017.04.001. 99

Bunn, F., Goodman, C., Sworn, K., Rait, G., Brayne, C., Robinson, L., McNeilly, E., and Iliffe, S. (2012). Psychosocial factors that shape patient and carer experiences of dementia diagnosis and treatment: a systematic review of qualitative studies. *PLOS Medicine*, 9(10), e1001331. DOI: 10.1371/journal.pmed.1001331. 71

Burkov, A. (2019). *The Hundred-Page Machine Learning Book*. Burkov, A. 169

Burleson, W., Lozano, C., Ravishankar, V., Lee, J., and Mahoney, D. (2018). An assistive technology system that provides personalized dressing support for people living with dementia: Capability study. *Journal of Medical Internet Research*, 6(2), e5587. DOI: 10.2196/medinform.5587. 87

Burmeister, O. K. (2016). The development of assistive dementia technology that accounts for the values of those affected by its use. *Ethics and Information Technology*, 18(3), 185–198. DOI: 10.1007/s10676-016-9404-2. 56, 71

Burningham, S. (2009). Developments in Canadian adult guardianship and co-decision-making law. *Dalhousie Journal of Legal Studies*, 18(1). https://digitalcommons.schulichlaw.dal.ca/djls/vol18/iss1/5. 53

Burr, J. A., Tavares, J., and Mutchler, J. E. (2011). Volunteering and hypertension risk in later life. *Journal of Aging and Health*, 23(1), 24–51. DOI: 10.1177/0898264310388272. 110

Business Wire. (2020). Juniper Research: Number of Voice Assistant Devices in Use to Overtake World Population by 2024, Reaching 8.4bn, Led by Smartphones. https://www.businesswire.com/news/home/20200427005609/en/Juniper-Research-Number-Voice-Assistant-Devices-Overtake.

Butchard, S. and Kinderman, P. (2019). Human rights, dementia, and identity. *European Psychologist*, 24(2), 159–168. DOI: 10.1027/1016-9040/a000370. 48

Caddell, L. S. and Clare, L. (2013). A profile of identity in early-stage dementia and a comparison with healthy older people. *Aging & Mental Health*, 17(3), 319–327. DOI: 10.1080/13607863.2012.742489. 36, 37

Caine, K. E., Zimmerman, C. Y., Schall-Zimmerman, Z., Hazlewood, W. R., Sulgrove, A. C., Camp, L. J., Connelly, K. H., Huber, L. L., and Shankar, K. (2010). DigiSwitch: Design and evaluation of a device for older adults to preserve privacy while monitoring health at home. *IHI'10 - Proceedings of the 1st ACM International Health Informatics Symposium*, 153–162. DOI: 10.1145/1882992.1883016. 77

Califf, R. M. and Muhlbaier, L. H. (2003). Health insurance portability and accountability act (HIPAA): Must there be a trade-off between privacy and quality of health care, or can we advance both? *Circulation*, 108(8), 915–918. DOI: 10.1161/01. CIR.0000085720.65685.90. 76

Callow, D. D., Arnold-Nedimala, N. A., Jordan, L. S., Pena, G. S., Won, J., Woodard, J. L., and Smith, J. C. (2020). The mental health benefits of physical activity in older adults survive the COVID-19 pandemic. *American Journal of Geriatric Psychiatry*, 28(10), 1046–1057. DOI: 10.1016/j.jagp.2020.06.024. 103

Camanio Care. (n.d.). BikeAround. https://www.camanio.com/us/products/bikearound/. 123

Campbell, N. L., Boustani, M. A., Skopelja, E. N., Gao, S., Unverzagt, F. W., and Murray, M. D. (2012). Medication adherence in older adults with cognitive impairment: A systematic evidence-based review. *The American Journal of Geriatric Pharmacotherapy*, 10(3), 165–177. DOI: 10.1016/j.amjopharm.2012.04.004. 87, 88

Canada's National Ballet School. (2021). Baycrest NBS sharing dance older adults. https://www. nbs-enb.ca/en/community-dance/baycrest-nbs-sharing-dance-older-adults. 132

Canadian Centre for Activity and Aging - Western University. (n.d.). Active at Home Videos. https://www.uwo.ca/ccaa/programs/videos/index.html. 103

Canadian Centre for Elder Law. (2019). *CCEL Report #10 | The Law and Practice of Health Care Consent for People Living with Dementia in British Columbia 2 Conversations about Care: The Law and Practice of Health Care Consent for People Living with Dementia in.* https:// www.bcli.org/wordpress/wp-content/uploads/2019/02/HCC_report-Final_web_Mar-29-2019.pdf. 59

Canadian Institute for Health Information. (2011). *Health Care in Canada, 2011: A Focus on Seniors and Aging.* https://www.homecareontario.ca/docs/default-source/publications-mo/ hcic_2011_seniors_report_en.pdf. 31

Canadian Institute for Health Information. (2020). Nursing in Canada, 2019. https://www.cihi.ca/ en/nursing-in-canada-2019. 11

Canadian Society for Exercise Physiology. (2020). Canadian 24-hour movement guidelines. https://csep.ca/guidelines. 103

Capgemini Research Institute. (2019). Smart Talk: How organizations and consumers are embracing voice and chat assistants. https://www.capgemini.com/wp-content/uploads/2019/09/Report---Conversational-Interfaces_Web-Final.pdf. 76

Carnemolla, P. (2018). Ageing in place and the internet of things – how smart home technologies, the built environment and caregiving intersect. *Visualization in Engineering*, 6(1), 1–16. DOI: 10.1186/s40327-018-0066-5. 165

CARP. (n.d.). Tories end forced retirement, decades of "age discrimination." https://www.carp.ca/2011/12/20/tories-end-forced-retirement-decades-of-age-discrimination/. 108

Carr, D. (2018). Volunteering among older adults: life dourse correlates and consequences. *The Journals of Gerontology: Series B*, 73(3), 479. DOI: 10.1093/geronb/gbx179. 110

Carr, D. C., Kail, B. L., Matz-Costa, C., and Shavit, Y. Z. (2018). Does becoming a volunteer attenuate loneliness among recently widowed older adults? *The Journals of Gerontology: Series B*, 73(3), 501–510. DOI: 10.1093/geronb/gbx092. 110

Carr, D.C. and Gunderson, J.A. (2016). The third age of life: Leveraging the mutual benefits of intergenerational engagement. *Public Policy & Aging Report*, 26(3), 83–87. DOI:10.1093/ppar/prw013. 121

Casey, B. (2019). The health of LGBTQIA2 communities in Canada. https://www.ourcommons.ca/Content/Committee/421/HESA/Reports/RP10574595/hesarp28/hesarp28-e.pdf. 11

Chang, J., Mcallister, C., and Mccaslin, R. (2015). Correlates of, and barriers to, internet use among older adults. *Journal of Gerontological Social Work*, 58(1), 66–85. DOI: 10.1080/01634372.2014.913754. 14, 98

Chang, P. J., Wray, L., and Lin, Y. (2014). Social relationships, leisure activity, and health in older adults. *Health Psychology*, 33(6), 516–523. DOI: 10.1037/hea0000051. 133

Charness, N. (2020). A framework for choosing technology interventions to promote successful longevity: Prevent, rehabilitate, augment, substitute (PRAS). *Gerontology*, 66(2), 169–175. DOI: 10.1159/000502141. 14

Charness, N., Kelley, C. L., Bosman, E. A., and Mottram, M. (2001). Word-processing training and retraining: Effects of adult age, experience, and interface. *Psychology and Aging*, 16(1), 110–127. DOI: 10.1037/0882-7974.16.1.110. 146

Chen, A. T., Ge, S., Cho, S., Teng, A. K., Chu, F., Demiris, G., and Zaslavsky, O. (2021). Reactions to COVID-19, information and technology use, and social connectedness among older adults with pre-frailty and frailty. *Geriatric Nursing*, 42(1), 188–195. DOI: 10.1016/j.gerinurse.2020.08.001. 98, 100

Chen, S.W. and Chippendale, T. (2018). Leisure as an end, not just a means, in occupational therapy intervention. *American Journal of Occupational Therapy*, 72(4), 7204347010p1–7204347010p5. DOI: 10.5014/ajot.2018.028316. 118

Chesham, A., Wyss, P., Müri, R. M., Mosimann, U. P., and Nef, T. (2017). What older people like to play: Genre preferences and acceptance of casual games. *JMIR Serious Games*, 5(2), e8. DOI: 10.2196/games.7025. 130

Cho, D., Post, J., and Kim, S. K. (2018). Comparison of passive and active leisure activities and life satisfaction with aging. *Geriatrics & Gerontology International*, 18(3), 380–386. DOI: 10.1111/ggi.13188. 119

Choi, N. and DiNitto, D. (2013a). The digital divide among low-income homebound older adults: Internet use patterns, eHealth literacy, and attitudes toward computer/Internet use. *Journal of Medical Internet Research*, 15(5), e93. DOI: 10.2196/jmir.2645. 112, 113

Choi, N. and DiNitto, D. (2013b). Internet use among older adults: association with health needs, psychological capital, and social capital. *Journal of Medical Internet Research*, 15(5), e97. DOI: 10.2196/jmir.2333. 14, 98, 133

Choi, N. G. and Bohman, T. M. (2007). Predicting the changes in depressive symptomatology in later life: How much do changes in health status, marital and caregiving status, work and volunteering, and health-related behaviors contribute? *Journal of Aging and Health*, 19(1), 152–177. DOI: 10.1177/0898264306297602. 110

Choi, Y., Demiris, G., and Thompson, H. (2018). Feasibility of smart speaker use to support aging in place. *Innovation in Aging*, 2(suppl_1), 560–560. DOI: 10.1093/geroni/igy023.2073. 93

Christman, J. and Anderson, J. (Eds.). (2005). *Autonomy and the Challenges to Liberalism: New Essays.* Cambridge: Cambridge University Press. DOI: 10.1017/CBO9780511610325. 43

Chung, J., Demiris, G., and Thompson, H. J. (2016). Ethical considerations regarding the use of smart home technologies for older adults: An integrative review. *Annual Review of Nursing Research*, 34(1), 155–181. DOI: 10.1891/0739-6686.34.155. 169

Clarke, C. L., Keady, J., Wilkinson, H., and Gibb, C. E. (2011). *Risk Assessment and Management for Living Well with Dementia*. London, UK: Jessica Kingsely Publishers. 62

Clarkson, J., Coleman, R., Keates, S., and Lebbon, C. (2013). From margins to mainstream. In *Inclusive Design*. Springer, pp. 1–25. DOI: 10.1007/978-1-4471-0001-0. 23

Classen, S., Shechtman, O., Awadzi, K. D., Joo, Y., and Lanford, D. N. (2010). Traffic violations versus driving errors of older adults: Informing clinical practice. *American Journal of Occupational Therapy*, 64(2), 233–241. DOI: 10.5014/ajot.64.2.233. 140

Collingwood, L. (2017). Privacy implications and liability issues of autonomous vehicles. *Information & Communications Technology Law*, 26(1), 32–45. DOI: 10.1080/13600834.2017.1269871. 141

Collins, J. M., Reizes, O., and Dempsey, M. K. (2016). Healthcare commercialization programs: Improving the efficiency of translating healthcare innovations from academia into practice. *IEEE Journal of Translational Engineering in Health and Medicine*, 4, 1–7. DOI: 10.1109/JTEHM.2016.2609915. 149

Conte, S. (2019). Design and evaluation of an interactive tactile aid to support older adults' adoption of technology. Master's thesis, University of Toronto. https://tspace.library.utoronto.ca/bitstream/1807/97958/3/Conte_Sho__201911_MIS_thesis.pdf. 99

Conte, S. and Munteanu, C. (2018). An interactive tactile aid for older adults learning to use tablet devices. *Extended Abstracts of the 2018 CHI Conference on Human Factors in Computing Systems*, 1–4. DOI: 10.1145/3170427.3186548. 99

Cook-Cottone, C. P. and Guyker, W. M. (2018). The development and validation of the Mindful Self-Care Scale (MSCS): An assessment of practices that support positive embodiment. *Mindfulness*, 9(1), 161–175. DOI: 10.1007/s12671-017-0759-1. 84

Cook, A. M., Polgar, J. M., and Encarnação, P. (2020). *Assistive Technologies: Principles and Practice*, 5th ed. St. Louis, MO. 83

Corregidor-Sánchez, A. I., Segura-Fragoso, A., Rodríguez-Hernández, M., Criado-Alvarez, J. J., González-Gonzalez, J., and Polonio-López, B. (2020). Can exergames contribute to improving walking capacity in older adults? A systematic review and meta-analysis. *Maturitas*, 132, 40–48. DOI: 10.1016/j.maturitas.2019.12.006. 104

Coşar, S., Fernandez-Carmona, M., Agrigoroaie, R., Pages, J., Ferland, F., Zhao, F., Yue, S., Bellotto, N., and Tapus, A. (2020). ENRICHME: Perception and interaction of an assistive robot for the elderly at home. *International Journal of Social Robotics*, 12(3), 779–805. DOI: 10.1007/s12369-019-00614-y. 127

Coughlan, G., Laczó, J., Hort, J., Minihane, A. M., and Hornberger, M. (2018). Spatial navigation deficits—Overlooked cognitive marker for preclinical Alzheimer disease? *Nature Reviews Neurology*, 14(8), 496–506. DOI: 10.1038/s41582-018-0031-x. 135

Creative Aging Calgary Society. (2021). Creative resources for seniors. http://www.creativeaging-calgary.com/blog/. 121

Crete-Nishihata, M., Baecker, R. M., Massimi, M., Ptak, D., Campigotto, R., Kaufman, L. D., Brickman, A. M., Turner, G. R., Steinerman, J. R., and Black, S. E. (2012). Reconstructing the past: Personal memory technologies are not just personal and not just for memory.

Human–Computer Interaction, 27(1–2), 92–123. DOI: 10.1080/07370024.2012.656062. 41, 129

Cruz-Jentoft, A. J., Bahat, G., Bauer, J., Boirie, Y., Bruyère, O., Cederholm, T., Cooper, C., Landi, F., Rolland, Y., Sayer, A. A., Schneider, S. M., Sieber, C. C., Topinkova, E., Vandewoude, M., Visser, M., Zamboni, M., Bautmans, I., Baeyens, J. P., Cesari, M., … Schols, J. (2019). Sarcopenia: Revised European consensus on definition and diagnosis. *Age and Ageing*, 48(1), 16–31. DOI: 10.1093/ageing/afy169. 104

Cruz-Sandoval, D., Favela, J., Lopez-Nava, I. H., and Morales, A. (2021). Adoption of wearable devices by persons with dementia: lessons from a non-pharmacological intervention enabled by a social robot. In *Studies in Computational Intelligence*, 933. Springer Science and Business Media Deutschland GmbH., pp. 145–163. DOI: 10.1007/978-981-15-9897-5_8. 147

Cuddy, A. J. C., Norton, M. I., and Fiske, S. T. (2005). This old stereotype: The pervasiveness and persistence of the elderly stereotype. *Journal of Social Issues*, 61(2), 267–285. DOI: 10.1111/j.1540-4560.2005.00405.x. 107

Cunningham, C., O'Sullivan, R., Caserotti, P., and Tully, M. A. (2020). Consequences of physical inactivity in older adults: A systematic review of reviews and meta-analyses. *Scandinavian Journal of Medicine & Science in Sports*, 30(5), 816–827. DOI: 10.1111/sms.13616. 102, 103

Czaja, S. J., Boot, W. R., Charness, N., and Rogers, W. A. (2019). *Designing for Older Adults: Principles and Creative Human Factors Approaches*, 3rd ed. CRC Press. 90, 108, 109, 119, 122, 125, 126, 130

Czaja, S. J. (2017). The potential role of technology in supporting older adults. *Public Policy & Aging Report*, 27(2), 44–48. DOI: 10.1093/ppar/prx006. 108, 109

Czarnuch, S. and Mihailidis, A. (2011). The design of intelligent in-home assistive technologies: Assessing the needs of older adults with dementia and their caregivers. *Gerontechnology*, 10(3), 169–182. DOI: 10.4017/gt.2011.10.3.005.00. 83, 175

Dale, B., Söderhamn, U., and Söderhamn, O. (2012). Life situation and identity among single older home-living people: A phenomenological-hermeneutic study. *International Journal of Qualitative Studies on Health & Well-being*, 7(1), 1–11. DOI: 10.3402/qhw.v7i0.18456. 83, 84

Davidson, J. and Schimmele, C. (2019). Study: Evolving internet use among Canadian seniors. In *Statistics Canada*. https://www150.statcan.gc.ca/n1/en/daily-quotidien/190710/dq190710d-eng.pdf?st=hBBUtqTO. 97, 165

Davies, H. (2015). Ted Cruz using firm that harvested data on millions of unwitting Facebook users. *The Guardian*. https://www.theguardian.com/us-news/2015/dec/11/senator-ted-cruz-president-campaign-facebook-user-data. 74

Davis, S., Nesbitt, K., and Nalivaiko, E. (2014). A systematic review of cybersickness. *Proceedings of the 2014 Conference on Interactive Entertainment*. DOI: 10.1145/2677758. 104

de Boer, M. E., Hertogh, C. M. P. M., Dröes, R. M., Jonker, C., and Eefsting, J. A. (2010). Advance directives in dementia: Issues of validity and effectiveness. *International Psychogeriatrics*, 22(2), 201–208. DOI: 10.1017/S1041610209990706. 58

De Courval, L. P., Gélinas, I., Gauthier, S., Gayton, D., Liu, L., Rossignol, M., Sampalis, J., and Dastoor, D. (2006). Reliability and validity of the Safety Assessment Scale for people with dementia living at home. *Canadian Journal of Occupational Therapy*, 73(2), 67–75. DOI: 10.1177/000841740607300201. 62

de Joode, E., Proot, I., Slegers, K., van Heugten, C., Verhey, F., and van Boxtel, M. (2012). The use of standard calendar software by individuals with acquired brain injury and cognitive complaints: A mixed methods study. *Disability and Rehabilitation: Assistive Technology*, 7(5), 389–398. DOI: 10.3109/17483107.2011.644623. 113

De Labra, C., Guimaraes-Pinheiro, C., Maseda, A., Lorenzo, T., and Millán-Calenti, J. C. (2015). Effects of physical exercise interventions in frail older adults: A systematic review of randomized controlled trials. *BMC Geriatrics*, 15(1), 1–16. DOI: 10.1186/s12877-015-0155-4. 103

de Vries, B. (2009). Aspects of life and death, grief and loss in lesbian, gay, bisexual andtransgender communities. In Doka, K. J. and Tucci, A. S., Eds., *Living with Grief: Diversity in End-of-Life Care*. Washington, DC: Hospice Foundation of America. 51

de Witte, L., Steel, E., Gupta, S., Ramos, V. D., and Roentgen, U. (2018). Assistive technology provision: Towards an international framework for assuring availability and accessibility of affordable high-quality assistive technology. *Disability and Rehabilitation: Assistive Technology*, 13(5), 467–472. DOI: 10.1080/17483107.2018.1470264. 26

Dementia Alliance International. (2016). The human rights of people living with dementia: From rhetoric to reality. https://www.dementiaallianceinternational.org/wp-content/uploads/2016/04/The-Human-Rights-of-People-Living-with-Dementia-from-Rhetoric-to-Reality.pdf. 53

Dementia Circle. (2017). About purple alert. 137, 138

Department for Constitutional Affairs. (2007). *Mental Capacity Act Code of Practice*. London: TSO. 43, 175

Desai, M. H. and McKinnon, B. J. (2020). Balance and dizziness disorders in the elderly: A review. *Current Otorhinolaryngology Reports*, 8(2), 198–207. DOI: 10.1007/s40136-020-00281-y. 72

DeSilva, J., Prensky-Pomeranz, R., and Zweig, M. (2021). Digital health consumer adoption report 2020. https://rockhealth.com/reports/digital-health-consumer-adoption-report-2020/. 171

Desmond, D., Layton, N., Bentley, J., Boot, F. H., Borg, J., Dhungana, B. M., Gallagher, P., Gitlow, L., Gowran, R. J., Groce, N., Mavrou, K., Mackeogh, T., McDonald, R., Pettersson, C., and Scherer, M. J. (2018). Assistive technology and people: A position paper from the first global research, innovation and education on assistive technology (GREAT) summit. *Disability and Rehabilitation: Assistive Technology*, 13(5), 437–444. DOI: 10.1080/17483107.2018.1471169. 25

Deterding, S., Dixon, D., Khaled, R., and Nacke, L. (2011). From game design elements to gamefulness: Defining "gamification." *Proceedings of the 15th International Academic MindTrek Conference: Envisioning Future Media Environments, MindTrek* 2011, 9–15. DOI: 10.1145/2181037.2181040. 65

Dewing, J. (2006). Wandering into the future: Reconceptualizing wandering "A natural and good thing." *International Journal of Older People Nursing*, 1(4), 239–249. DOI: 10.1111/j.1748-3743.2006.00045.x. 68

DiGennaro Reed, F. D., Strouse, M. C., Jenkins, S. R., Price, J., Henley, A. J., and Hirst, J. M. (2014). Barriers to independent living for individuals with disabilities and seniors. *Behavior Analysis in Practice*, 7(2), 70–77. DOI: 10.1007/S40617-014-0011-6. xxi

Digital Canada. (2019). ACE: Building localized self-sovereign identity ecosystems. https://digitalcanada.io/ace-ssi/. 54

Dollard, J., Barton, C., Newbury, J., and Turnbull, D. (2012). Falls in old age: A threat to identity. *Journal of Clinical Nursing*, 21(17–18), 2617–2625. DOI: 10.1111/j.1365-2702.2011.03990.x. 35, 38

Donath, L., Rössler, R., and Faude, O. (2016). Effects of virtual reality training (exergaming) compared to alternative exercise training and passive control on standing balance and functional mobility in healthy community-dwelling seniors: A meta-analytical review. *Sports Medicine*, 46(9), 1293–1309. DOI: 10.1007/s40279-016-0485-1. 103, 174

Doron, I. (2003). Mental incapacity, guardianship, and the elderly: An exploratory study of Ontario's consent and capacity board. *Canadian Journal of Law and Society*, 18(1), 131–145. DOI: 10.1017/S082932010000750X. 52

Douglas, M. (1990). Risk as a forensic resource. *Daedalus*, 119(4), 1–16. http://www.jstor.org/stable/20025335. 63

Dove, E. S., Kelly, S. E., Lucivero, F., Machirori, M., Dheensa, S., and Prainsack, B. (2017). Beyond individualism: Is there a place for relational autonomy in clinical practice and research? *Clinical Ethics*, 12(3), 150–165. DOI: 10.1177/1477750917704156. 12

Dryden, J. (2014). Autonomy. *Internet Encyclopedia of Philosophy*. https://iep.utm.edu/autonomy/. 5, 173

Dulcey-Ruiz, E. (2015). *Envejecimiento y Vejez Categorias y Conceptos*. Bogotá, Colombia: Fundación Cepsiger para el desarrollo humano, p. 588. 12

Dworkin, G. (2015). The nature of autonomy. *Nordic Journal of Studies in Educational Policy*, 2015(2), 28479. DOI: 10.3402/nstep.v1.28479. 5, 160, 173

Dworkin, R. (1993). Life's Dominion: An Argument about Abortion, Euthanasia, and Individual Freedom. New York, NY: Vintage Books, a Division of Random House, Inc. 39

Eberly, L. A., Kallan, M. J., Julien, H. M., Haynes, N., Khatana, S. A. M., Nathan, A. S., Snider, C., Chokshi, N. P., Eneanya, N. D., Takvorian, S. U., Anastos-Wallen, R., Chaiyachati, K., Ambrose, M., O'Quinn, R., Seigerman, M., Goldberg, L. R., Leri, D., Choi, K., Gitelman, Y., ... Adusumalli, S. (2020). Patient characteristics associated with telemedicine access for primary and specialty ambulatory care during the COVID-19 pandemic. *JAMA Network Open*, 3(12), 2031640. DOI: 10.1001/jamanetworkopen.2020.31640. 147

Eby, D. W., Molnar, L. J., Zhang, L., St. Louis, R. M., Zanier, N., Kostyniuk, L. P., and Stanciu, S. (2016). Use, perceptions, and benefits of automotive technologies among aging drivers. *Injury Epidemiology*, 3(1), 28. DOI: 10.1186/s40621-016-0093-4. 139

Eccles, M. P. and Mittman, B. S. (2006). Welcome to implementation science. *Implementation Science*, 1(1). DOI: 10.1186/1748-5908-1-1. 151, 175

Echt, K. V. and Burridge, A. B. (2011). Predictors of reported internet use in older adults with high and low health literacy: The role of socio-demographics and visual and cognitive function. *Physical & Occupational Therapy In Geriatrics*, 29(1), 23–43. DOI: 10.3109/02703181.2010.547657. 98

Egan, M. Y., Laliberte Rudman, D., Ceci, C., Kessler, D., McGrath, C., Gardner, P., King, J., Lanoix, M., and Malhotra, R. (2017). Seniors, risk and rehabilitation: Broadening our thinking.

Disability and Rehabilitation, 39(13), 1348–1355. DOI: 10.1080/09638288.2016.1192227. 64

Eichler, T., Hoffmann, W., Hertel, J., Richter, S., Wucherer, D., Michalowsky, B., Dreier, A., and Thyrian, J. R. (2016). Living alone with dementia: Prevalence, correlates and the utilization of health and nursing care services. *Journal of Alzheimer's Disease*, 52(2), 619–629. DOI: 10.3233/JAD-151058. 69

Encarnação, P. and Cook, A. (2017). Fundamentals of robotic assistive technologies. In Encarnação, P., Ed., *Robotic Assistive Technologies*. New York, NY: CRC Press. pp. 1–24. 23

Ennis, A., Rafferty, J., Synnott, J., Cleland, I., Nugent, C., Selby, A., McIlroy, S., Berthelot, A., and Masci, G. (2017). A smart cabinet and voice assistant to support independence in older adults. *International Conference on Ubiquitous Computing and Ambient Intelligence*, 10586, 466–472. DOI: 10.1007/978-3-319-67585-5_47. 93

Entwistle, V. A., Carter, S. M., Cribb, A., and McCaffery, K. (2010). Supporting patient autonomy: The importance of clinician-patient relationships. *Journal of General Internal Medicine*, 25(7), 741–745. DOI: 10.1007/s11606-010-1292-2. 7, 160

European Commission Directorate General Environment. (2009). Bridging the Valley of Death: Public support for commercialisation of eco-innovation. ec.europa.eu/environment/enveco/rd_innovation/pdf/studies/bridging_valley_report.pdf. 150

Everitt, D. E., Fields, D. R., Soumerai, S. S., and Avorn, J. (1991). Resident behavior and staff distress in the nursing home. *Journal of the American Geriatrics Society*, 39(8), 792–798. DOI: 10.1111/j.1532-5415.1991.tb02702.x. 63

Faber, K. and Lierop, D. (2020). How will older adults use automated vehicles? Assessing the role of AVs in overcoming perceived mobility barriers. *Transportation Research Part A: Policy and Practice*, 133, 353–363. DOI: 10.1016/j.tra.2020.01.022. 139, 141

Fadillah, N. and Ihsan, A. (2020). Smart bed using voice recognition for paralyzed patient. *IOP Conference Series: Materials Science and Engineering*, 854(1), 012045. DOI: 10.1088/1757-899X/854/1/012045. 93

FamliNet. (2021). App for Seniors - FamliNet. https://www.famlinet.com/. 82

Fang, M. L., Canham, S. L., Battersby, L., Sixsmith, J., Wada, M., and Sixsmith, A. (2018). Exploring privilege in the digital divide: implications for theory, policy, and practice. *The Gerontologist*, 59(1), E1–E15. DOI: 10.1093/geront/gny037. 14, 27, 174

Farivar, S., Abouzahra, M., and Ghasemaghaei, M. (2020). Wearable device adoption among older adults: A mixed-methods study. *International Journal of Information Management*, 55, 102209. DOI: 10.1016/j.ijinfomgt.2020.102209. 100, 101

Fazio, S., Pace, D., Flinner, J., and Kallmyer, B. (2018). The fundamentals of person-centered care for individuals with dementia. *The Gerontologist*, 58(suppl_1), S10–S19. DOI: 10.1093/geront/gnx122. 38

Ferguson, C., Hickman, L. D., Turkmani, S., Breen, P., Gargiulo, G., and Inglis, S. C. (2021). "Wearables only work on patients that wear them": Barriers and facilitators to the adoption of wearable cardiac monitoring technologies. *Cardiovascular Digital Health Journal*, 2(2), 137–147. DOI: 10.1016/j.cvdhj.2021.02.001. 147

Ferguson, W. K. (2014). A policy framed analysis of The Valley Of Death in U.S. university technology. Doctoral dissertation, The University of Southern Mississippi. https://aquila.usm.edu/dissertations. 149

Fernandes, A. L. and Colello, K. J. (2009). Alert systems for missing adults in eleven states: Background and issues for Congress. https://www.policyarchive.org/handle/10207/18546. 138

Fernandez-Cervantes, V., Neubauer, N., Hunter, B., Stroulia, E., and Liu, L. (2018). VirtualGym: A kinect-based system for seniors exercising at home. *Entertainment Computing*, 27, 60–72. DOI: 10.1016/j.entcom.2018.04.001. 105

Ferreira, J. S. S. P., Sacco, I. C. N., Siqueira, A. A., Almeida, M. H. M., and Sartor, C. D. (2019). Rehabilitation technology for self-care: Customised foot and ankle exercise software for people with diabetes. *PLOS ONE*, 14(6), e0218560. DOI: 10.1371/journal.pone.0218560. 99

Fetherstonhaugh, D., Rayner, J. A., and Tarzia, L. (2019). Hanging on to some autonomy in decisionmaking: How do spouse carers support this? *Dementia*, 18(4), 1219–1236. DOI: 10.1177/1471301216678104. 72

Field, T. (2018). Online dating profiles and problems in older adults: A review. *OBM Geriatrics*, 2(3), 012. DOI: 10.21926/obm.geriatr.1803012. 126

Fijal, D. and Beagan, B. L. (2019). Indigenous perspectives on health: Integration with a Canadian model of practice. *Canadian Journal of Occupational Therapy*, 86(3), 220–231. DOI: 10.1177/0008417419832284. 161, 162

Filsinger, M. and Freitag, M. (2019). Internet use and volunteering: Relationships and differences across age and applications. *Voluntas*, 30(1), 87–97. DOI: 10.1007/s11266-018-0045-4. 112

Finkelstein, E. A., Haaland, B. A., Bilger, M., Sahasranaman, A., Sloan, R. A., Nang, E. E. K., and Evenson, K. R. (2016). Effectiveness of activity trackers with and without incentives to increase physical activity (TRIPPA): A randomised controlled trial. *The Lancet Diabetes and Endocrinology*, 4(12), 983–995. DOI: 10.1016/S2213-8587(16)30284-4. 100

Fleetwood, J. (2017). Public health, ethics, and autonomous vehicles. *American Journal of Public Health*, 107(4), 532–537. DOI: 10.2105/AJPH.2016.303628. 139, 140, 141

Floegel, T. A., Florez-Pregonero, A., Hekler, E. B., and Buman, M. P. (2017). Validation of consumer-based hip and wrist activity monitors in older adults with varied ambulatory abilities. *The Journals of Gerontology: Series A*, 72(2), 229–236. DOI: 10.1093/GERONA/GLW098. 100

Folstein, M., Folstein, S., and McHugh, P. (1975). "Mini-mental state." A practical method for grading the cognitive state of patients for the clinician. *Journal of Psychiatric Research*, 12(3), 189–198. DOI: 10.1016/0022-3956(75)90026-6. 44

Fong, J. H., Mitchell, O. S., and Koh, B. S. K. (2015). Disaggregating activities of daily living limitations for predicting nursing home admission. *Health Services Research*, 50(2), 560–578. DOI: 10.1111/1475-6773.12235. 86

Formosa, M. (2012). Education and older adults at the university of the Third Age. *Educational Gerontology*, 38(2), 114–126. DOI: 10.1080/03601277.2010.515910. 121

Frank, C., Sink, C., Mynatt, L., Rogers, R., and Rappazzo, A. (1996). Surviving the "Valley of Death": A comparative analysis. *Journal of Technology Transfer* (21), 61–69. DOI: 10.1007/BF02220308. 149

Freiesleben, S. D., Megges, H., Herrmann, C., Wessel, L., and Peters, O. (2021). Overcoming barriers to the adoption of locating technologies in dementia care: A multi-stakeholder focus group study. *BMC Geriatrics*, 21(1), 1–17. DOI: 10.1186/S12877-021-02323-6. 64

Frik, A., Nurgalieva, L., Bernd, J., Lee, J. S., Schaub, F., and Egelman, S. (2019). Privacy and security threat models and mitigation strategies of older adults. *Proceedings of the 15th Symposium on Usable Privacy and Security*, SOUPS 2019, 21–40. 73, 74, 76, 77, 78

Fristedt, S., Svärdh, S., Löfqvist, C., Schmidt, S. M., and Iwarsson, S. (2021). "Am I representative (of my age)? No, I'm not"—Attitudes to technologies and technology development differ but unite individuals across rather than within generations. *PLOS ONE*, 16(4), e0250425. DOI: 10.1371/JOURNAL.PONE.0250425. 40

Future of Today Institute. (2021). Tech Trends Reports. 14th Annual Edition. https://futuretoday-institute.com/trends/. 92, 101, 102

Ganyo, M., Dunn, M., and Hope, T. (2011). Ethical issues in the use of fall detectors. *Ageing and Society*, 31(8), 1350–1367. DOI: 10.1017/S0144686X10001443. 78

Gao, Z., Lee, J. E., McDonough, D. J., and Albers, C. (2020). Virtual reality exercise as a coping strategy for health and wellness promotion in older adults during the COVID-19 pandemic. *Journal of Clinical Medicine*, 9(6), 1986. DOI: 10.3390/jcm9061986. 104

Gavett, B. E., Zhao, R., John, S. E., Bussell, C. A., Roberts, J. R., and Yue, C. (2017). Phishing suspiciousness in older and younger adults: The role of executive functioning. *PLOS ONE*, 12(2), e0171620. DOI: 10.1371/journal.pone.0171620. 100

Gell, N. M., Rosenberg, D. E., Demiris, G., LaCroix, A. Z., and Patel, K. V. (2015). Patterns of technology use among older adults with and without disabilities. *The Gerontologist*, 55(3), 412–421. DOI: 10.1093/geront/gnt166. 99

Gelonch, O., Ribera, M., Codern-Bové, N., Ramos, S., Quintana, M., Chico, G., Cerulla, N., Lafarga, P., Radeva, P., and Garolera, M. (2019). Acceptability of a lifelogging wearable camera in older adults with mild cognitive impairment: A mixed-method study. *BMC Geriatrics*, 19(1), 1–10. DOI: 10.1186/s12877-019-1132-0. 128, 176

Gergerich, E. and Davis, L. (2017). Silver alerts: A notification system for communities with missing adults. *Journal of Gerontological Social Work*, 60(3), 232–244. DOI: 10.1080/01634372.2017.1293757. 138

Gilroy, R. (2008). Places that support human flourishing: Lessons from later life. *Planning Theory & Practice*, 9(2), 145–163. DOI: 10.1080/1464935080204154. 110

Gitlow, L. (2014). Technology use by older adults and barriers to using technology. *Physical & Occupational Therapy In Geriatrics*, 32(3), 271–280. DOI: 10.3109/02703181.2014.946640. 99

Gomes, M., Figueiredo, D., Teixeira, L., Poveda, V., Paúl, C., Santos-Silva, A., and Costa, E. (2017). Physical inactivity among older adults across Europe based on the SHARE database. *Age and Ageing*, 46(1), 71–77. DOI: 10.1093/ageing/afw165. 103

Gonçalves-Bradley, D. C., Iliffe, S., Doll, H. A., Broad, J., Gladman, J., Langhorne, P., Richards, S. H., and Shepperd, S. (2017). Early discharge hospital at home. *Cochrane Database of Systematic Reviews*, 2017(6). DOI: 10.1002/14651858.CD000356.pub4. 104

Gonzales, E., Matz-Costa, C., and Morrow-Howell, N. (2015). Increasing opportunities for the productive engagement of older adults: A response to population aging. *The Gerontologist*, 55(2), 252–261. DOI: 10.1093/geront/gnu176. 110

Goodall, N. J. (2016). Can you program ethics into a self-driving car? *IEEE Spectrum*, 53(6). DOI: 10.1109/MSPEC.2016.7473149. 141

Göransson, C., Wengström, Y., Ziegert, K., Langius-Eklöf, A., Eriksson, I., Kihlgren, A., and Blomberg, K. (2017). Perspectives of health and self-care among older persons—To be implemented in an interactive information and communication technology-platform. *Journal of Clinical Nursing*, 26(23–24), 4745–4755. DOI: 10.1111/jocn.13827. 84, 98

Gordon, N. P. and Hornbrook, M. C. (2018). Older adults' readiness to engage with eHealth patient education and self-care resources: A cross-sectional survey. *BMC Health Services Research*, 18(1), 1–13. DOI: 10.1186/s12913-018-2986-0. 100

Gotcare.ca. (2018). The need for an alternative home care delivery mechanism. Toronto, Ontario: Gotcare. 47

Gotcare. (2021). Gotcare. https://gotcare.ca/. 47

Gottlieb, B. H. and Gillespie, A. A. (2008). Volunteerism, health, and civic engagement among older adults. *Canadian Journal on Aging / La Revue Canadienne Du Vieillissement*, 27(4), 399–406. https://muse.jhu.edu/article/266019. 110

Gouvernement du Quebec. (2020). Issue my directives in case of incapacity | Régie de l'assurance maladie du Québec (RAMQ). https://www.ramq.gouv.qc.ca/en/citizens/health-insurance/issue-directives-case-incapacity. 57

Government of Alberta. (2021a). About capacity assessment. https://www.alberta.ca/capacity-assessment.aspx. 44

Government of Alberta. (2021b). Adult guardianship. https://www.alberta.ca/adult-guardianship.aspx. 52

Government of Alberta. (2021c). Trusteeship. https://www.alberta.ca/trusteeship.aspx. 53

Government of Canada. (2016). What every older Canadian should know about: Powers of attorney (for financial matters and property) and joint bank accounts. https://www.canada.ca/en/employment-social-development/corporate/seniors/forum/power-attorney-financial.html. 52, 176

Gow, A. J., Pattie, A., and Deary, I. J. (2017). Lifecourse activity participation from early, mid, and later adulthood as determinants of cognitive aging: The Lothian Birth Cohort 1921. *The Journals of Gerontology Series B: Psychological Sciences and Social Sciences*, 72(1), 25–37. DOI: 10.1093/geronb/gbw124. 118

Graham, S., Depp, C., Lee, E. E., Nebeker, C., Tu, X., Kim, H. C., and Jeste, D. V. (2019). Artificial intelligence for mental health and mental illnesses: An overview. *Current Psychiatry Reports*, 21(11), 1–18. DOI: 10.1007/s11920-019-1094-0. 131

Greenhalgh, T., Robert, G., Macfarlane, F., Bate, P., and Kyriakidou, O. (2004). Diffusion of innovations in service organizations: Systematic review and recommendations. *Milbank Quarterly*, 82(4), 581–629. DOI: 10.1111/j.0887-378X.2004.00325.x. 149

Greenhalgh, T., Wherton, J., Papoutsi, C., Lynch, J., Hughes, G., A'Court, C., Hinder, S., Fahy, N., Procter, R., and Shaw, S. (2017). Beyond adoption: A new framework for theorizing and evaluating nonadoption, abandonment, and challenges to the scale-up, spread, and sustainability of health and care technologies. *Journal of Medical Internet Research*, 19(11), e8775. DOI: 10.2196/jmir.8775. 146

Grindrod, K. A., Li, M., and Gates, A. (2014). Evaluating user perceptions of mobile medication management applications with older adults: A usability study. *JMIR Mhealth and Uhealth*, 2(1), e11. DOI: 10.2196/mhealth.3048. 88

Grotz, J., Dyson, S., and Birt, L. (2020). Pandemic policy making: The health and wellbeing effects of the cessation of volunteering because of restrictions during the COVID-19 pandemic on older adults. *Quality in Ageing and Older Adults*, 21(4), 261–269. DOI: 10.1108/QAOA-07-2020-0032. 110, 112

Grübler, A. (2003). *Technology and Global Change*. Cambridge, U.K.: Cambridge University Press, p. 464. 17, 177

Guiney, H., Keall, M., and Machado, L. (2021). Volunteering in older adulthood is associated with activity engagement and cognitive functioning. *Aging, Neuropsychology, and Cognition*, 28(2), 253–269. DOI: 10.1080/13825585.2020.1743230. 110

Gutman, G. M., Kwon, S., Güttler, J. F., Georgoulas, C., Linner, T., Bock, T., Fukuda, R., and Kwon, S. (2016). Smart home technologies supporting aging in place. In Kwon, S., Ed., *Gerontechnology*. New York, NY: Springer Publishing Company, pp. 223–249. 91

Güttler, J., Bittner, A., Langosch, K., Bock, T., and Mitsukura, Y. (2018). Development of an affordable and easy-to-install fall detection system. *IEEJ Transactions on Electrical and Electronic Engineering*, 13(5), 664–670. DOI: 10.1002/TEE.22648. 64, 78

Haddad, L. M. and Geiger, R. A. (2020). *Nursing Ethical Considerations*. Treasure Island, FL: StatPearls Publishing. 69

Hall, S., Dodd, R. H., and Higginson, I. J. (2014). Maintaining dignity for residents of care homes: A qualitative study of the views of care home staff, community nurses, residents and their families. *Geriatric Nursing*, 35(1), 55–60. DOI: 10.1016/j.gerinurse.2013.10.012. 74

Halvorsen, C. J. and Morrow-Howell, N. (2017). A conceptual framework on self-employment in later life: Toward a research agenda. *Work, Aging and Retirement*, 3(4), 313–324. DOI: 10.1093/workar/waw031. 108

Hamel, L. W. (2017). Views and experiences with end-of-life medical care in the U.S. Kaiser Family Foundation. https://www.kff.org/other/report/views-andexperiences-with-end-of-life-medical-care-in-the-u-s/. 51

Hammell, K. R. W. (2014). Belonging, occupation, and human well-being: An exploration. *Canadian Journal of Occupational Therapy*, 81(1), 39–50. DOI: 10.1177/0008417413520489. 162

Hammell, K. W. (2009). Sacred texts: A sceptical exploration of the assumptions underpinning theories of occupation. *Canadian Journal of Occupational Therapy*, 76(1), 6–22. DOI: 10.1177/000841740907600105. 12, 118

Hammell, K. W. (2011). Resisting theoretical imperialism in the disciplines of occupational science and occupational therapy. *British Journal of Occupational Therapy*, 74(1), 27–33. DOI: 10.4276/030802211X12947686093602. 162

Harding, R. (2012). Legal constructions of dementia: discourses of autonomy at the margins of capacity. *Journal of Social Welfare and Family Law*, 34(4), 425–442. DOI: 10.1080/09649069.2012.755031. 43, 48, 50

Hargittai, E., Piper, A. M., and Morris, M. R. (2019). From internet access to internet skills: Digital inequality among older adults. *Universal Access in the Information Society*, 18(4), 881–890. DOI: 10.1007/s10209-018-0617-5. 14, 98

Harris, K., Krygsman, S., Waschenko, J., and Laliberte Rudman, D. (2017). Ageism and the older worker: A scoping review. *The Gerontologist*, 58(2), e1–e14. DOI: 10.1093/geront/gnw194. 109

Harris, P. S., Payne, L., Morrison, L., Green, S. M., Ghio, D., Hallett, C., Parsons, E. L., Aveyard, P., Roberts, H. C., Sutcliffe, M., Robinson, S., Slodkowska-Barabasz, J., Little, P. S., Stroud, M. A., and Yardley, L. (2019). Barriers and facilitators to screening and treating malnutrition in older adults living in the community: A mixed-methods synthesis. *BMC Family Practice*, 20(1), 1–10. DOI: 10.1186/s12875-019-0983-y. 89

Havens, E., Slabaugh, S. L., Helmick, C. G., Cordier, T., Zack, M., Gopal, V., and Prewitt, T. (2017). Comorbid arthritis is associated with lower health-related quality of life in older adults with other chronic conditions, United States, 2013–2014. *Preventing Chronic Disease*, 14(7), 160495. DOI: 10.5888/pcd14.160495. 85

Hawryluck, L. (n.d.). Estate planning: Ten practical steps to improve written advance directives in powers of attorney for healthcare. *McGill Journal of Law and Health*. https://mjlh.mcgill.ca/estate-planning/estate-planning-ten-practical-steps-to-improve-written-advance-directives-in-powers-of-attorney-for-healthcare-9/. 57

Haynes, E. M. K., Neubauer, N. A., Cornett, K. M. D., O'connor, B. P., Jones, G. R., and Jakobi, J. M. (2020). Age and sex-related decline of muscle strength across the adult lifespan: A scoping review of aggregated data. *Applied Physiology, Nutrition and Metabolism*, 45(11), pp. 1185–1196. DOI: 10.1139/apnm-2020-0081. 104

Hazel, M. (2018). Reasons for working at 60 and beyond. *Statistics Canada*. https://www150.statcan.gc.ca/n1/pub/71-222-x/71-222-x2018003-eng.htm. 107

Health Law Institute. (n.d.). Advance directives. http://eol.law.dal.ca/?page_id=231. 51

Hebblethwaite, S. (2017). The (in)visibility of older adults in digital leisure cultures. In Carnicelli, S., McGillivray, D., and McPherson, G., Eds., *Digital Leisure Cultures: Critical Perspectives*. London, England: Taylor & Francis, pp. 94–106. 126

Héder, M. (2017). From NASA to EU: The evolution of the TRL scale in public sector innovation. *The Innovation Journal*, 22(2), 1–23. http://eprints.sztaki.hu/9204/. 149

Hellström, Y. and Hallberg, I. R. (2004). Determinants and characteristics of help provision for elderly people living at home and in relation to quality of life. *Scandinavian Journal of Caring Sciences*, 18(4), 387–395. DOI: 10.1111/j.1471-6712.2004.00291.x. 83

Hensel, B. K., Demiris, G., and Courtney, K. L. (2006). Defining obtrusiveness in home telehealth technologies: A conceptual framework. *Journal of the American Medical Informatics Association*, 13(4), 428–431. DOI: 10.1197/jamia.m2026. 169

Hevelke, A. and Nida-Rümelin, J. (2014). Responsibility for crashes of autonomous vehicles: An ethical analysis. *Science and Engineering Ethics*, 21(3), 619–630. DOI: 10.1007/S11948-014-9565-5. 141

Hilbert, M. (2011). The end justifies the definition: The manifold outlooks on the digital divide and their practical usefulness for policy-making. *Telecommunications Policy*, 35(8), 715–736. DOI: 10.1016/j.telpol.2011.06.012. 14, 27, 174

Hodge, H., Carson, D., Carson, D., Newman, L., and Garrett, J. (2017). Using Internet technologies in rural communities to access services: The views of older people and service providers. *Journal of Rural Studies*, 54, 469–478. DOI: 10.1016/j.jrurstud.2016.06.016. 133

Hodges, S., Berry, E., and Wood, K. (2011). SenseCam: A wearable camera that stimulates and rehabilitates autobiographical memory. *Memory*, 19(7), 685–696. DOI: 10.1080/09658211.2011.605591. 41

Hollinda, K., Ferguson-King, C., Unrau, B., Daum, C., Ríos-Rincón, A., and Liu, L. (2019). Occupational therapists as digital storytelling facilitators: Exploring roles and strategies. *CAOT Conference 2019 | Congrès de l'ACE*. https://www.caot.ca/document/7338/Conference%202019%20Program.pdf. 42

Holmes, C. (1996). Surrogate decision making in the 90s: Learning to respect our Elders. *University of Toledo Law Review*, 28. 53

Hootman, J. M., Helmick, C. G., and Brady, T. J. (2012). A public health approach to addressing arthritis in older adults: The most common cause of disability. *American Journal of Public Health*, 102(3), 426–433. DOI: 10.2105/AJPH.2011.300423. 85

Hovbrandt, P., Fridlund, B., and Carlsson, G. (2007). Very old people's experience of occupational performance outside the home: Possibilities and limitations. *Scandinavian Journal of Occupational Therapy*, 14(2), 77–85. DOI: 10.1080/11038120600773013. 110

Høy, B., Wagner, L., and Hall, E. O. C. (2007). Self-care as a health resource of elders: An integrative review of the concept. *Scandinavian Journal of Caring Sciences*, 21(4), 456–466. DOI: 10.1111/j.1471-6712.2006.00491.x. 83

Hu, R., Linner, T., Trummer, J., Güttler, J., Kabouteh, A., Langosch, K., and Bock, T. (2020). Developing a smart home solution based on personalized intelligent interior units to promote activity and customized healthcare for aging society. *Journal of Population Ageing*, 13(2), 257–280. DOI: 10.1007/S12062-020-09267-6. 106

Hughes, J. C. and Baldwin, C. (2006). *Ethical Issues in Dementia Care: Making Difficult Decisions*. Jessica Kingsley Publishers. 64, 174

Hunsaker, A. and Hargittai, E. (2018). A review of Internet use among older adults. *New Media & Society*, 20(10), 3937–3954. DOI: 10.1177/1461444818787348. 14, 97, 98, 99, 126, 165

Hunt, M. R. and Ells, C. (2011). Partners towards autonomy: Risky choices and relational autonomy in rehabilitation care. *Disability and Rehabilitation*, 33(11), 961–967. DOI: 10.3109/09638288.2010.515703. 70

Hunter, I. M. and Lockhart, C. (2020). Use of sensor data to support interRAI assessments of older adults living at home in New Zealand. Palmerston North: Massey University, 2020 for NZ Ministry of Health. 147

Hunter, I., Phoebe, E., Lockhart, C., Guesgen, H., Whiddett, D., and Singh, A. (2021). Telehealth at home: Co-designing a smart home telehealth system. In Maeder, A.J., Higa, C., Van Den Berg, M.E.L., and Gough, C, *Telehealth Innovations in Remote Healthcare Services Delivery Global Telehealth*, IOS Press, pp. 47–56. 146, 147

Ibrahim, J. E., Holmes, A., Young, C., and Bugeja, L. (2019). Managing risk for aging patients in long-term care: A narrative review of practices to support communication, documentation, and safe patient care practices. *Risk Management and Healthcare Policy*, 12, 31–39. DOI: 10.2147/RMHP.S159073. 70

Ibrahim, N. I. and Davies, S. (2012). Aging: Physical difficulties and safety in cooking tasks. *Work*, 41, 5152–5159. DOI: 10.3233/WOR-2012-0804-5152. 88

Insight Timer. (2021). Insight timer. https://insighttimer.com/. 129

International Consortium on Dementia and Wayfinding. (n.d.). *International Consortium on Dementia and Wayfinding*. https://icdw.org/. 69

International Labour Organization. (2015). World employment and social outlook. http://www.ilo.org/wcmsp5/groups/public/---dgreports/---dcomm/---publ/documents/publication/wcms_337069.pdf. 107

International Organization for Standardization. (2007). Medical devices—application of risk management to medical devices (Patent No. ISO 14971:2007). https://www.iso.org/standard/38193.html. 61, 62, 64, 175, 176, 177

Jacobs, K. and Simon, L. (2015). *Quick Reference Dictionary For Occupational Therapy. In Occupational Therapy In Health Care*, 6th ed. Slack Incorporated. DOI: 10.3109/07380577.2016.1141349. 83, 173

Jadczyk, T., Kiwic, O., Khandwalla, R. M., Grabowski, K., Rudawski, S., Magaczewski, P., Benyahia, H., Wojakowski, W., and Henry, T. D. (2019). Feasibility of a voice-enabled automated platform for medical data collection: CardioCube. *International Journal of Medical Informatics*, 129, 388–393. DOI: 10.1016/j.ijmedinf.2019.07.001. 177

James, J. B., Morrow-Howell, N., Gonzales, E., Matz-Costa, C., and Riddle-Wilder, A. (2020). Beyond the livelong workday: Is there a new face of retirement? In Czaja, S. J. Sharit, J., and James, J. B., Eds., *Current and Emerging Trends in Aging and Work*. Cham, Switzerland: Springer International Publishing. DOI: 10.1007/978-3-030-24135-3_18. 108

Jamieson, M., Cooper Reed, A., Amaral, E., and Cameron, J. I. (2021). Exploring the emergence of self-directed home care in Ontario: A qualitative case study on Gotcare services. *Home Health Care Management & Practice*, 33(1), 28–36. DOI: 10.1177/1084822320953840. 47

Janke, M. C., Nimrod, G., and Kleiber, D. A. (2008). Leisure activity and depressive symptoms of widowed and married women in later life. *Journal of Leisure Research*, 40(2), 250–266. DOI: 10.1080/00222216.2008.11950140. 133

Jaworska, A. (2017). Ethical dilemmas in neurodegenerative disease: Respecting patients at the twilight of agency. In Illes, J., Ed., *Neuroethics: Anticipating the Future*. Oxford University Press. pp. 273–293. DOI: 10.1093/oso/9780198786832.003.0015. 48

Jones, S. and Fox, S. (2009). Generations online in 2009. In *Pew Internet & American Life Project*. https://www.pewresearch.org/internet/2009/01/28/generations-online-in-2009/. 97

Jopp, D. S., Wozniak, D., Damarin, A. K., De Feo, M., Jung, S., and Jeswani, S. (2015). How could lay perspectives on successful aging complement scientific theory? Findings from a US and a German life-span sample. *The Gerontologist*, 55(1), 91-106. DOI: 10.1093/geront/gnu059. 79

Jovanović, M., De Angeli, A., McNeill, A., and Coventry, L. (2021). User requirements for inclusive technology for older adults. *International Journal of Human–Computer Interaction*, 1–19. DOI: 10.1080/10447318.2021.1921365. 146, 147

Joy for All. (2018). Lifelike robotic pets for seniors. Ageless Innovation LLC. https://joyforall.com/. 128.

Jung, H. T., Daneault, J. F., Lee, H., Kim, K., Kim, B., Park, S., Ryu, T., Kim, Y., and Ivan Lee, S. (2019). Remote assessment of cognitive impairment level based on serious mobile game performance: An initial proof of concept. *IEEE Journal of Biomedical and Health Informatics*, 23(3), 1269–1277. DOI: 10.1109/JBHI.2019.2893897. 45

Kadylak, T., Blocker, K., Ramadhani, W., Koon, L., Khaleghi, R., Kovac, C., Sreenivas, R., and Rogers, W. (2020). Developing digital home assistant user guides for older adults with and without long-term mobility disabilities. *Innovation in Aging*, 4(Supplement 1), 645–646. DOI: 10.1093/geroni/igaa057.2220. 93, 141

Kamimura, T., Ishiwata, R., and Inoue, T. (2012). Medication reminder device for the elderly patients with mild cognitive impairment. *American Journal of Alzheimer's Disease and Other Dementias*, 27(4), 238–242. DOI: /10.1177/1533317512450066. 88

Kappen, D. L., Mirza-Babaei, P., and Nacke, L. E. (2019). Older adults' physical activity and exergames: A systematic review. *International Journal of Human–Computer Interaction*, 35(2), 140–167. DOI: 10.1080/10447318.2018.1441253. 103, 174

Kaufman, D., Ma, M., Sauvé, L., Renaud, L., and Dupláa, E. (2019). Benefits of digital gameplay for older adults: Does game type make a difference? *International Journal of Aging Research*, 2(4), 43. DOI: 10.28933/ijoar-2019-07-2805. 132

Kautonen, T., Down, S., and South, L. (2008). Enterprise support for older entrepreneurs: The case of PRIME in the UK. *International Journal of Entrepreneurial Behaviour & Research*, 14(2), 85–101. DOI: 10.1108/13552550810863071. 152

Kawagley, A. O. and Barnhardt, R. (1999). A long journey: Alaska onward to excellence in Yupiit/Tuluksak schools. Case study. https://eric.ed.gov/?id=ED437254. 13

Kawate, I. (2021). China's young and old rail against raising retirement age - Nikkei Asia. Nikkei Asia. https://asia.nikkei.com/Spotlight/Society/China-s-young-and-old-rail-against-raising-retirement-age. 107

Kekade, S., Hseieh, C. H., Islam, M. M., Atique, S., Mohammed Khalfan, A., Li, Y. C., and Abdul, S. S. (2018). The usefulness and actual use of wearable devices among the elderly population. *Computer Methods and Programs in Biomedicine*, 153, 137–159. DOI: 10.1016/j.cmpb.2017.10.008. 147

Kerkhof, Y., Pelgrum-Keurhorst, M., Mangiaracina, F., Bergsma, A., Vrauwdeunt, G., Graff, M., and Dröes, R.-M. (2019). User-participatory development of FindMyApps: A tool to help people with mild dementia find supportive apps for self-management and meaningful activities. *Digital Health*, 5, 2055207618822942. DOI: 10.1177/2055207618822942. 129

Keshavarz, B., Ramkhalawansingh, R., Haycock, B., Shahab, S., and Campos, J. L. (2018). Comparing simulator sickness in younger and older adults during simulated driving under different multisensory conditions. *Transportation Research Part F: Traffic Psychology and Behaviour*, 54, 47–62. DOI: 10.1016/J.TRF.2018.01.007. 104

Khasnabis, C., Mirza, Z., and MacLachlan, M. (2015). Opening the GATE to inclusion for people with disabilities. *The Lancet*, 386(10010), 2229–2230. DOI: 10.1016/S0140-6736(15)01093-4. 21, 22, 24, 25, 26

Kintsakis, A. M., Reppou, S. E., Karagiannis, G. T., and Mitkas, P. A. (2015). Robot-assisted cognitive exercise in mild cognitive impairment patients: The RAPP approach. *2015 E-Health and Bioengineering Conference* (EHB), 1–4. DOI: 10.1109/EHB.2015.7391355. 127

Kirby, A. V. (2015). Beyond independence: Introducing deweyan philosophy to the dialogue on occupation and independence. *Journal of Occupational Science*, 22(1), 17–25. DOI: 10.1080/14427591.2013.803297. 7

Kivimäki, T., Stolt, M., Charalambous, A., and Suhonen, R. (2020). Safety of older people at home: An integrative literature review. *International Journal of Older People Nursing*, 15(1), 15. DOI: 10.1111/opn.12285. 88

Klimczuk, A. (2016). Creative aging: Drawing on the arts to enhance healthy aging. In Pachana, N. A., Ed., *Encyclopedia of Geropsychology*. Springer Singapore, pp. 1–5. DOI: 10.1007/978-981-287-080-3_363-1. 132

Kline, S. J. (1985). What is technology? *Bulletin of Science, Technology & Society*, 5(3), 215–218. DOI: 10.1177/027046768500500301. 17, 177

Knight-Davidson, P., Lane, P., and McVicar, A. (2020). Methods for co-creating with older adults in living laboratories: A scoping review. *Health and Technology* 2020 10:5, 10(5), 997–1009. DOI: 10.1007/S12553-020-00441-6. 147

Knoefel, F., Wallace, B., Goubran, R., Sabra, I., and Marshall, S. (2019). Semi-autonomous vehicles as a cognitive assistive device for older adults. *Geriatrics*, 4(4), 63. DOI: 10.3390/geriatrics4040063. 139

Knox, W. B. and Stone, P. (2009). Interactively shaping agents via human reinforcement: The TAMER framework. *K-CAP'09 - Proceedings of the 5th International Conference on Knowledge Capture*, 9–16. DOI: 10.1145/1597735.1597738. 131

Koenig, H. G. (2012). Religion, spirituality, and health: The research and clinical implications. *International Scholarly Research Notices*, 2012, 1–33. DOI: 10.5402/2012/278730. 130

Kohl, H. W., Craig, C. L., Lambert, E. V., Inoue, S., Alkandari, J. R., Leetongin, G., Kahlmeier, S., Andersen, L. B., Bauman, A. E., Blair, S. N., Brownson, R. C., Bull, F. C., Ekelund, U., Goenka, S., Guthold, R., Hallal, P. C., Haskell, W. L., Heath, G. W., Katzmarzyk, P. T., ... Wells, J. C. (2012). The pandemic of physical inactivity: Global action for public health. *The Lancet*, 380(9838), 294–305. DOI: 10.1016/S0140-6736(12)60898-8. 103

Kohon, J. and Carder, P. (2014). Exploring identity and aging: Auto-photography and narratives of low income older adults. *Journal of Aging Studies*, 30, 47–55. DOI: 10.1016/j.jaging.2014.02.006. 35

Koo, S. H., Lee, Y. Bin, Kim, C., Kim, G., Lee, G., and Koh, J. S. (2020). Development of gait assistive clothing-typed soft wearable robot for elderly adults. *International Journal of Clothing Science and Technology*, 33(4). DOI: 10.1108/IJCST-04-2020-0052. 124

Kottorp, A., Nygård, L., Hedman, A., Öhman, A., Malinowsky, C., Rosenberg, L., Lindqvist, E., and Ryd, C. (2016). Access to and use of everyday technology among older people: An occupational justice issue – but for whom? *Journal of Occupational Science*, 23(3), 382–388. DOI: 10.1080/14427591.2016.1151457. 27, 99, 165

Kroemeke, A. (2015). Skala postrzeganej autonomii: Struktura czynnikowa i własciwosci psychometryczne polskiej adaptacji [Perceived Autonomy in Old Age scale: Factor structure and psychometric properties of the Polish adaptation]. *Psychiatria Polska*, 49(1), 107–117. DOI: 10.12740/PP/22616. 31

Kuykendall, L., Tay, L., and Ng, V. (2015). Leisure engagement and subjective well-being: A meta-analysis. *Psychological Bulletin*, 141(2), 364–403. DOI: 10.1037/a0038508. 118

Kwok, T. C. Y., Yuen, K. S. L., Ho, F. K. Y., and Chan, W. M. (2010). Getting lost in the community: A phone survey on the community-dwelling demented people in Hong Kong. *International Journal of Geriatric Psychiatry*, 25(4), 427–432. DOI: 10.1002/gps.2361. 135

L'Engle, M. (2001). *Walking on Water: Reflections on Faith and Art*. WaterBrook Press. 31

Lachance, E. L. (2020). COVID-19 and its impact on volunteering: Moving towards virtual volunteering. *Leisure Sciences*, 43, 1–7. DOI: 10.1080/01490400.2020.1773990. 112

Lancioni, G., Singh, N., Sigafoos, J., Pinto, K., De Vanna, F., and Caffò, A. (2017). A technology-aided program for helping persons with Alzheimer's disease perform daily activities. *Journal of Enabling Technologies*, 11(3), 85–91. DOI: 10.1108/JET-03-2017-0011. 90

Lancioni, G., VivianaPerilli, Singh, N. N., O'Reilly, M. F., Sigafoos, J., Cassano, G., Pinto, K., Minervini, M. G., and Oliva, D. (2012). Technology-aided pictorial cues to support the performance of daily activities by persons with moderate Alzheimer's disease. *Research in Developmental Disabilities*, 33(1), 265–273. DOI: 10.1016/j.ridd.2011.09.017. 90

Laposha, I. and Smallfield, S. (2020). Examining the occupational therapy definition of self-care: A scoping review. *Occupational Therapy In Health Care*, 34(2), 99–115. DOI: 10.1080/07380577.2019.1703238. 83, 84

Larkin, M. (2018). BikeAround: Making memories accessible and world travel feasible for all. *The Journal on Active Aging*, 17(5), 76–80. www.icaa.cc. 123

Lashua, B., Johnson, C. W., and Parry, D. C. (2021). Leisure in the time of coronavirus: A rapid response special issue. *Leisure Sciences*, 43(1–2), 6–11. DOI: 10.1080/01490400.2020.1774827. 118

Laufer, Y., Dar, G., and Kodesh, E. (2014). Does a Wii-based exercise program enhance balance control of independently functioning older adults? A systematic review. *Clinical Interventions in Aging*, 9, 1803–1813. DOI: 10.2147/CIA.S69673. 104

Lauzé, M., Martel, D. D., Agnoux, A., Sirois, M. J., Émond, M., Daoust, R., and Aubertin-Leheudre, M. (2018). Feasibility, acceptability and effects of a home-based exercise program using a gerontechnology on physical capacities after a minor injury in community-living older adults: A pilot study. *Journal of Nutrition, Health and Aging*, 22(1), 16–25. DOI: 10.1007/s12603-017-0938-8. 104

Lavretsky, H. (2010). Spirituality and aging. *Aging Health*, 6(6), 749–769. DOI: 10.2217/ahe.10.70. 129

Law Commission of Ontario. (2020). "Legal capacity": Setting the standard. https://www.lco-cdo.org/en/capacity-guardianship-discussion-paper-partII-sectionI. 43, 47

Le, B. M., Impett, E. A., Lemay, E. P., Muise, A., and Tskhay, K. O. (2018). Communal motivation and well-being in interpersonal relationships: An integrative review and meta-analysis. *Psychological Bulletin*, 144(1), 1–25. DOI: 10.1037/bul0000133. 9

Lee. (2019). Older adults' digital gameplay, social capital, social connectedness, and civic participation. *Game Studies*, 19(1). http://gamestudies.org/1901/articles/lee. 132

Lee, B. C., Xie, J., Ajisafe, T., and Kim, S. H. (2020). How are wearable activity trackers adopted in older adults? Comparison between subjective adoption attitudes and physical activity performance. *International Journal of Environmental Research and Public Health*, 17(10), 3461. DOI: 10.3390/ijerph17103461. 101

Leger, S. J., Dean, J. L., Edge, S., and Casello, J. M. (2019). "If I had a regular bicycle, I wouldn't be out riding anymore": Perspectives on the potential of e-bikes to support active living and independent mobility among older adults in Waterloo, Canada. *Transportation Research Part A: Policy and Practice*, 123, 240–254. DOI: 10.1016/j.tra.2018.10.009. 123

Leist, A. K. (2013). Social media use of older adults: A mini-review. *Gerontology*, 59(4), 378–384. DOI: 10.1159/000346818. 126

Lesakova, D. (2016). Seniors and their food shopping behavior: An empirical analysis. *Social and Behavioral Sciences*, 220, 243–250. DOI: 10.1016/j.sbspro.2016.05.496. 90

Leuty, V., Boger, J., Young, L., Hoey, J., and Mihailidis, A. (2013). Engaging older adults with dementia in creative occupations using artificially intelligent assistive technology. *Assistive Technology*, 25(2), 72–79. DOI: 10.1080/10400435.2012.715113. 132

Leyland, L. A. (2016). E-bikes and their benefit for older adults. *International Conference on Traffic and Transport Psychology* (ICTTP). https://www.cycleboom.org/wp-content/uploads/2016/08/cB_ICTTP_Aug16.pdf. 123

Leyland, L. A., Spencer, B., Beale, N., Jones, T., and van Reekum, C. M. (2019). The effect of cycling on cognitive function and well-being in older adults. *PLOS ONE*, 14(2), e0211779. DOI: 10.1371/journal.pone.0211779. 123

LG Newsroom. (2020). LG announces autonomous robot with disinfecting UV light for various B2B applications. https://www.lgnewsroom.com/2020/12/lg-announces-autonomous-robot-with-disinfecting-uv-light-for-various-b2b-applications/. 92

Li-Hua, R. (2012). Definitions of technology. In Olsen, J. K. B., Pedersen, S. A., and Hendricks, V. F., Eds., *A Companion to the Philosophy of Technology*. Wiley-Blackwell. pp. 18–22. 17, 177

Li, J. and Porock, D. (2014). Resident outcomes of person-centered care in long-term care: A narrative review of interventional research. *International Journal of Nursing Studies*, 51(10), pp. 1395–1415. Pergamon. DOI: 10.1016/j.ijnurstu.2014.04.003. 38

Lifetime Arts. (2020). Older adult choir instructional videos from SFCMC. https://www.lifetimearts.org/blog/covid-19_resource/older-adult-choir-instructional-videos-from-sfcmc/. 121

Lignell, A. (2014). Older consumers' adoption of online shopping. Master's thesis, Lappeenranta University of Technology. https://lutpub.lut.fi/bitstream/handle/10024/95639/

older-consumers-adoption-of-online-shopping-lignell-annamari.pdf?sequence=3&is-Allowed=y. 90

Lin, C.-C. C., Dievler, A., Robbins, C., Sripipatana, A., Quinn, M., and Nair, S. (2018). Telehealth in health centers: Key adoption factors, barriers, and opportunities. *Health Affairs*, 37(12), 1967–1974. DOI: 10.1377/hlthaff.2018.05125. 171

Liu, L. (2018). Occupational therapy in the Fourth Industrial Revolution. *Canadian Journal of Occupational Therapy*, 85(4), 272–285. DOI: 10.1177/0008417418815179. 17, 84

Liu, L., Miguel-Cruz, A., and Ríos-Rincón, A. M. (2019). Technology acceptance, adoption, and usability: Arriving at consistent terminologies and measurement approaches. In Hayre, C.M., Muller, D.J.; and Scherer, M.J., Eds., *Everyday Technologies in Healthcare*. Boca Raton, FL: CRC Press, pp. 319–338. DOI: 10.1201/9781351032186. 145, 149

Liu, L., Miguel Cruz, A., and Juzwishin, D. (2018). Caregivers as a proxy for responses of dementia clients in a GPS technology acceptance study. *Behaviour and Information Technology*, 37(6), 634–645. DOI: 10.1080/0144929X.2018.1470672. 148, 160

Liu, L., Miguel Cruz, A., Rios Rincon, A., Buttar, V., Ranson, Q., and Goertzen, D. (2015). What factors determine therapists' acceptance of new technologies for rehabilitation-a study using the Unified Theory of Acceptance and Use of Technology (UTAUT). *Disability and Rehabilitation*, 37(5), 447–455. DOI: 10.3109/09638288.2014.923529. 160

Liu, L., Miguel Cruz, A., Ruptash, T., Barnard, S., and Juzwishin, D. (2017). Acceptance of Global Positioning System (GPS) technology among dementia clients and family caregivers. *Journal of Technology in Human Services*, 35(2), 99–119. DOI: 10.1080/15228835.2016.1266724. 148

Liu, L., Stroulia, E., Nikolaidis, I., Miguel-Cruz, A., and Rios Rincon, A. (2016). Smart homes and home health monitoring technologies for older adults: A systematic review. *International Journal of Medical Informatics*, 91, 44–59. DOI: 10.1016/j.ijmedinf.2016.04.007. 91, 92, 102

Löckenhoff, C. E., De Fruyt, F., Terracciano, A., McCrae, R. R., De Bolle, M., Costa, P. T., Aguilar-Vafaie, M. E., Ahn, C. kyu, Ahn, H. nie, Alcalay, L., Allik, J., Avdeyeva, T. V., Barbaranelli, C., Benet-Martinez, V., Blatný, M., Bratko, D., Cain, T. R., Crawford, J. T., Lima, M. P., ... Yik, M. (2009). Perceptions of aging across 26 cultures and their culture-level associates. *Psychology and Aging*, 24(4), 941–954. DOI: 10.1037/a0016901. 107

Loeb, S. J. (2006). African American older adults coping with chronic health conditions. *Journal of Transcultural Nursing*, 17(2), 139–147. DOI: 10.1177/1043659605285415. 84

Lucchetti, G., Vitorino, L. M., Nasri, F., and Lucchetti, A. L. G. (2019). Impact of religion and spirituality in older persons. In Lucchetti, G., Prieto Peres, M., and Damiano, R., Eds., *Spirituality, Religiousness and Health*. 4, pp. 115–130. Springer, Cham. DOI: 10.1007/978-3-030-21221-6_8. 129

Lundgren, B. (2020). Safety requirements vs. crashing ethically: What matters most for policies on autonomous vehicles. *AI & Society*, 36(2), 405–415. DOI: 10.1007/S00146-020-00964-6. 141

Luoma-Halkola, H. and Häikiö, L. (2020). Independent living with mobility restrictions: Older people's perceptions of their out-of-home mobility. *Ageing and Society*, 1–22. DOI: 10.1017/S0144686X20000823. 10

Lupton, D. (1993). Risk as moral danger: The social and political functions of risk discourse in public health. *International Journal of Health Services*, 23(3), 425–435. DOI: 10.2190/16AY-E2GC-DFLD-51X2. 63

Lupton, D. (1999). Risk and Sociocultural Theory: New Directions and Perspectives. Lupton, D., Ed. Cambridge University Press. DOI: 10.1017/CBO9780511520778. 63

Lupton, D. (2005). Risk as moral danger: The social and political functions of risk discourse in public health. In P. Conrad (Ed.), *The Sociology of Health & Illness: Critical Perspectives* (pp. 422–431). Worth Publishers. 63, 176

Maâlaoui, A., Castellano, S., Safraou, I., and Bourguiba, M. (2013). An exploratory study of seniorpreneurs: A new model of entrepreneurial intentions in the French context. *International Journal of Entrepreneurship and Small Business*, 20(2), 148–164. DOI: 10.1504/IJESB.2013.056276. 108

Macdonald, A., and Dening, T. (2002). Dementia is being avoided in NHS and social care. *BMJ*, 324(7336), 548a – 548. DOI: 10.1136/bmj.324.7336.548/a. 63

MacNeil, M., Koch, M., Kuspinar, A., Juzwishin, D., Lehoux, P., and Stolee, P. (2019). Enabling health technology innovation in Canada: Barriers and facilitators in policy and regulatory processes. *Health Policy*, 123(2), 203–214. DOI: 10.1016/j.healthpol.2018.09.018. 149

Makita, M., Mas-Bleda, A., Stuart, E., and Thelwall, M. (2021). Ageing, old age and older adults: A social media analysis of dominant topics and discourses. *Ageing and Society*, 41(2), 247–272. DOI: 10.1017/S0144686X19001016. 107

Malone, J. and Dadswell, A. (2018). The role of religion, spirituality and/or belief in positive ageing for older adults. *Geriatrics*, 3(2), 28. DOI: 10.3390/geriatrics3020028. 129, 130

Man-Son-Hing, M., Marshall, S. C., Molnar, F. J., and Wilson, K. G. (2007). Systematic review of driving risk and the efficacy of compensatory strategies in persons with dementia. *Journal of the American Geriatrics Society*, 55(6), 878–884. DOI: 10.1111/j.1532-5415.2007.01177.x. 63

Mankins, J. C. (1995). Technology readiness levels for renewable energy sectors. http://www.artemisinnovation.com/images/TRL_White_Paper_2004-Edited.pdf. 149

Manning, L. K. (2013). Navigating hardships in old age: Exploring the relationship between spirituality and resilience in later life. *Qualitative Health Research*, 23(4), 568–575. DOI: 10.1177/1049732312471730. 129

Manthorpe, J. (2003). Community mental health nursing and dementia care: Practice perspectives. In Keady, J., Clarke, C. L., and Adams, T., Eds., *Community Mental Health Nursing And Dementia Care: Practice Perspectives*. Philadelphia, PA: Open University Press, p. 63. 63

Mario. (2015). Mario Project. http://www.mario-project.eu/portal/. 127

Marston, H. R. and Kowert, R. (2020). What role can videogames play in the COVID-19 pandemic? *Emerald Open Research*, 2, 34. DOI: 10.35241/emeraldopenres.13727.2. 126, 132

Marston, H. R., Niles-Yokum, K., Earle, S., Gomez, B., and Lee, D. M. (2020). OK Cupid, stop bumbling around and match me tinder: Using dating apps across the life course. *Gerontology and Geriatric Medicine*, 6, 2333721420947498. DOI: 10.1177/2333721420947498. 126

Martel, D., Lauzé, M., Agnoux, A., Fruteau de Laclos, L., Daoust, R., Émond, M., Sirois, M. J., and Aubertin-Leheudre, M. (2018). Comparing the effects of a home-based exercise program using a gerontechnology to a community-based group exercise program on functional capacities in older adults after a minor injury. *Experimental Gerontology*, 108, 41–47. DOI: 10.1016/j.exger.2018.03.016. 104

Martinson, M. and Minkler, M. (2006). Civic engagement and older adults: A critical perspective. *The Gerontologist*, 46(3), 318–324. DOI: 10.1093/geront/46.3.318. 111

Massimi, M., Berry, E., Browne, G., Smyth, G., Watson, P., and Baecker, R. M. (2008). An exploratory case study of the impact of ambient biographical displays on identity in a patient with Alzheimer's disease. *Neuropsychological Rehabilitation*, 18(5–6), 742–765. DOI: 10.1080/09602010802130924. 41

MatchWork. (2020). MatchWork a finalist in the AGE-WELL National Impact Challenge. https://mymatchwork.com/blog/matchwork-a-finalist-in-the-age-well-national-impact-challenge-2/. 109

Mattson, D. C. (2015). Usability assessment of a mobile app for art therapy. *Arts in Psychotherapy*, 43, 1–6. DOI: 10.1016/j.aip.2015.02.005. 132

Matuska, K. M. and Christiansen, C. (2011). Ways of Living (Fourth ed.). AOTA Press. 83, 173

Mccreadie, C. and Tinker, A. (2005). The acceptability of assistive technology to older people. *Aging & Society*, 25(1), pp. 91–110. DOI: 10.1017/S0144686X0400248X. 39

McDarby, M., Llaneza, D., George, L., and Kozlov, E. (2020). Mobile applications for advance care planning: A comprehensive review of features, quality, content, and readability. *American Journal of Hospice and Palliative Medicine*, 38(8), pp. 983–994. DOI: 10.1177/1049909120959057. 57

McGrath, C., Molinaro, M. L., Sheldrake, E. J., Laliberte Rudman, D., and Astell, A. (2019). A protocol paper on the preservation of identity: Understanding the technology adoption patterns of older adults with age-related vision loss (ARVL). *International Journal of Qualitative Methods*, 18, 160940691983183. DOI: 10.1177/1609406919831833. 40

McNeill, A., Briggs, P., Pywell, J., and Coventry, L. (2017). Functional privacy concerns of older adults about pervasive health-monitoring systems. *ACM International Conference Proceeding Series*, Part F1285, 96–102. DOI: 10.1145/3056540.3056559. 73, 176

Meek, P. D. (2014). Resident and patient elopements: An overview of legal issues and trends. *Journal of Legal Nurse Consulting*, 25(2), 18–21. 70

Mehta, S. S. (2008). *Commercializing Successful Biomedical Technologies: Basic Principles for the Development of Drugs, Diagnostics and Devices*. Cambridge: Cambridge University Press. DOI: 10.1017/CBO9780511791345. 148

Meltzer, E. P., Kapoor, A., Fogel, J., Elbulok-Charcape, M. M., Roth, R. M., Katz, M. J., Lipton, R. B., and Rabin, L. A. (2017). Association of psychological, cognitive, and functional variables with self-reported executive functioning in a sample of nondemented community-dwelling older adults. *Applied Neuropsychology:Adult*, 24(4), pp. 364–375. DOI: h10.1080/23279095.2016.1185428. 45

Menghi, R., Gullà, F., and Germani, M. (2018). Assessment of a Smart Kitchen to Help People with Alzheimer's Disease. In Mokhtari, M., Abdulrazak, B., and Aloulou, H., Eds., *Smart Homes and Health Telematics, Designing a Better Future: Urban Assisted Living*. Springer International Publishing. pp. 304–309. DOI: 10.1007/978-3-319-94523-1_30. 90

Meyer, M. H. (2020). *Grandmothers at Work during Coronavirus*. https://surface.syr.edu/cgi/viewcontent.cgi?article=1052&context=lerner&preview_mode=1&z=1612816787. 114

Mihailidis, A., Boger, J., Hoey, J., and Jiancaro, T. (2011). Zero effort technologies: Considerations, challenges, and use in health, wellness, and rehabilitation. *Synthesis Lectures on Assistive, Rehabilitative, and Health-Preserving Technologies*, 1(2), 1–94. DOI: 10.2200/s00380ed-1v01y201108arh002. 149

Mihailidis, A., Boger, J. N., Craig, T., and Hoey, J. (2008). The COACH prompting system to assist older adults with dementia through handwashing: An efficacy study. *BMC Geriatrics*, 8(1), 1–18. DOI: 10.1186/1471-2318-8-28. 87

Mihailidis, A. and Sixsmith, A. (2021). Creating research products. In Mihailidis, A., Ed., *Knowledge, Innovation, and Impact* (1st ed.). Springer, Cham, pp. 265–267. DOI: 10.1007/978-3-030-34390-3. 148

Mimamoriai Project. (2021). Mimamoriai. https://mimamoriai.net/#appandsticker. 138

Mitroff, S. and Price, M. (2020). 6 ways seniors can use Google Home to make the COVID-19 quarantine easier. *Cnet*. https://www.cnet.com/home/smart-home/6-ways-seniors-can-use-google-home-to-make-the-covid-19-quarantine-easier/. 93

Mitzner, T. L., Chen, T. L., Kemp, C. C., Rogers, W. A., Mitzner, T. L., Chen, T. L., Kemp, C. C., and Rogers, W. A. (2014). Identifying the potential for robotics to assist older adults in different living environments. *International Journal of Social Robotics*, 6(2), 213–227. DOI: 10.1007/s12369-013-0218-7. 87

Mitzner, T. L., Savla, J., Boot, W. R., Sharit, J., Charness, N., Czaja, S. J., and Rogers, W. A. (2019). Technology adoption by older adults: Findings from the PRISM trial. *The Gerontologist*, 59(1), 34–44. DOI: 10.1093/geront/gny113. 113

Morrow-Howell, N., Galucia, N., and Swinford, E. (2020). Recovering from the COVID-19 pandemic: A focus on older adults. *Journal of Aging & Social Policy*, 32(4–5), 526–535. DOI: 10.1080/08959420.2020.1759758. 84, 90, 165

Motitech. (n.d.). Motitech. https://motitech.ca/. 124

Mück, J. E., Ünal, B., Butt, H., and Yetisen, A. K. (2019). Market and patent analyses of wearables in medicine. *Trends in Biotechnology*, 37(6), 563–566. DOI: 10.1016/j.tibtech.2019.02.001. 103

Mühle, A., Grüner, A., Gayvoronskaya, T., and Meinel, C. (2018). A survey on essential components of a self-sovereign identity. *Computer Science Review*, 30, 80–86. DOI: 10.1016/j.cosrev.2018.10.002. 54

Mukherjee, D. (2011). Participation of older adults in virtual volunteering: A qualitative analysis. *Ageing International*, 36(2), 253–266. DOI: 10.1007/s12126-010-9088-6. 111

Mulders, J. O. (2019). Employers' age-related norms, stereotypes and ageist preferences in employment. *International Journal of Manpower*, 41(5), pp. 523–534. DOI: 10.1108/IJM-10-2018-0358. 109

Müller, I., Mertin, M., and M.A., R. (2017). Technology as an area of conflict between autonomy and safety - acceptance and attitudes of family caregivers in regard to technical assistance to ensure safe areas of movement for people with dementia diseases. *Proceedings of the 3rd International Conference on Information and Communication Technologies for Ageing Well and E-Health*, 2, 127–134. DOI: 10.5220/0006283001270134. 72

Mulligan, J. F. and Levin, S. M. (2000). Litigating nursing home wandering cases. *Elder's Advisor*, 2. https://heinonline.org/HOL/Page?handle=hein.journals/marqelad2&id=110&div=&collection=. 70

Muraco, A. (2006). Intentional families: Fictive kin ties between cross-gender, different sexual orientation friends. *Journal of Marriage and Family*, 68(5), 1313–1325. DOI: 10.1111/j.1741-3737.2006.00330.x. 51

Muse, E. D., Barrett, P. M., Steinhubl, S. R., and Topol, E. J. (2017). Towards a smart medical home. *The Lancet*, 389(10067), 358. DOI: 10.1016/S0140-6736(17)30154-X. 91

Musical Health Technologies. (2021). *SingFit*. https://www.singfit.com/. 132

Naffine, N. M. (2005). The legal presumption of reason: Noble truth, useful fiction, ignoble lie. *Cleveland State Law Review*, 53, 1–10. 43, 44

Nasreddine, Z. S., Phillips, N. A., Bédirian, V., Charbonneau, S., Whitehead, V., Collin, I., Cummings, J. L., and Chertkow, H. (2005). The Montreal Cognitive Assessment, MoCA: A brief screening tool for mild cognitive impairment. *Journal of the American Geriatrics Society*, 53, 695–699. DOI: 10.1111/j.1532-5415.2005.53221.x. 44

National Highway Traffic Safety Administration. (2017). NHTSA. NHTSA; U.S. Department of Transportation. https://www.nhtsa.gov/technology-innovation/automated-vehicles-safety. 140

National Research Council. (2007). SBIR and the Phase III Challenge of Commercialization. In *SBIR and the Phase III Challenge of Commercialization*. National Academies Press. DOI: 10.17226/11851. 150

National Seniors Strategy. (n.d.). Ensuring older Canadians and their caregivers are enabled to participate in informed decision- making and advance care planning. https://nationalseniorsstrategy.ca/the-four-pillars/pillar-2/advance-care-planning/. 57

Nayton, K., Fielding, E., Brooks, D., Graham, F. A., and Beattie, E. (2014). Development of an education program to improve care of patients with dementia in an acute care setting. *Journal of Continuing Education in Nursing*, 45(12), 552–558. DOI: 10.3928/00220124-20141023-04. 69

Neal, D. P., Kerkhof, Y. J. F., Ettema, T. P., Muller, M., Bosmans, J., Finnema, E., Graff, M., Dijkstra, K., Stek, M. L., and Dröes, R. M. (2021). Evaluation of FindMyApps: Protocol for a randomized controlled trial of the effectiveness and cost-effectiveness of a tablet-based intervention to improve self-management and social participation of community-dwelling people with mild dementia, compared to usual table use. *BMC Geriatrics*, 21(1), 138. DOI: 10.1186/s12877-021-02038-8. 129

Nedelsky, J. (1989). Reconceiving autonomy: Sources, thoughts, possibilities. *Yale Journal of Law & Feminism*, 1, 7–36. 43, 176

Neisser, U. (2008). Five kinds of self-knowledge. *Philosophical Psychology*, 1(1), 35–59. DOI: 10.1080/09515088808572924. 35

Neubauer, N. (2019). A framework to describe the levels of risk associated with dementia-related wandering. Doctoral dissertation, University of Alberta. https://era.library.ualberta.ca/items/ca2f0151-cf67-4bfe-980a-907de00630ed/view/9cfcb9f1-4ecf-4e60-b25f-53645882bf77/Neubauer_Noelannah_201908_PhD.pdf. 69

Neubauer, N. A. and Liu, L. (2021). Influence of perspectives on user adoption of wander-management strategies. *Dementia*, 20(2), 734–758. DOI: 10.1177/1471301220911304. 48

Newcomb, B. (2020). COVID-19: Why it's time to talk about advance directives. Leonard Davis School of Gerontology. https://gero.usc.edu/2020/04/03/covid-19-and-advance-directives/. 51

Nielsen, J. (1993). *Usability Engineering*. San Francisco, CA: Morgan Kaufmann. 145, 146, 177

Niemeijer, A. R., Depla, M., Frederiks, B., Francke, A. L., and Hertogh, C. (2014). CE: Original research the use of surveillance technology in residential facilities for people with dementia or intellectual disabilities. A study among nurses and support staff. *American Journal of Nursing*, 114(12), 28–37. DOI: 10.1097/01.NAJ.0000457408.38222.d0. 70

Nienaber, M. (2021). Germany's Scholz rejects further hike of retirement age to 68, Reuters. https://www.usnews.com/news/world/articles/2021-06-08/germanys-scholz-rejects-further-hike-of-retirement-age-to-68. 107

Nimrod, G., Janke, M. C., Gibson, H., and Singleton, J. (2012). Leisure across the later lifespan. In Gibson, H. J. and Singleton, J. F., Eds., *Leisure and Aging: Theory and Practice*. Human Kinetics Publishers Inc, pp. 95–109. 119

Nimrod, G. (2009). The internet as a resource in older adult leisure. *International Journal of Disability and Human Development*, 8(3), 207–214. DOI: 10.1515/IJDHD.2009.8.3.207. 133

Norman, C. D. and Skinner, H. A. (2006). eHEALS: The eHealth literacy scale. *Journal of Medical Internet Research*, 8(4), e507. DOI: 10.2196/jmir.8.4.e27. 97

NU CEPAL. (2017). Derechos de las personas mayores: Retos para la interdependencia y autonomía. https://repositorio.cepal.org/handle/11362/41471. 12

Nuzum, H., Stickel, A., Corona, M., Zeller, M., Melrose, R. J., and Wilkins, S. S. (2020). Potential benefits of physical activity in MCI and dementia. *Behavioural Neurology*, 2020. DOI: 10.1155/2020/7807856. 103

O'Brien, K., Liggett, A., Ramirez-Zohfeld, V., Sunkara, P., and Lindquist, L. A. (2020). Voice-controlled intelligent personal assistants to support aging in place. *Journal of the American Geriatrics Society*, 68(1), 176–179. DOI: 10.1111/jgs.16217. 93, 177

O'Neil, C. (2016). *Weapons of Math Destruction: How Big Data Increases Inequality and Threatens*, 1st ed. Crown. 170

OECD. (2017). Public Procurement for Innovation: Good Practices and Strategies, OECD Public Governance Reviews. *OECD*. DOI: 10.1787/9789264265820-en. 150

Olson, D. M. (2017). Caregiver or care-partner. *Journal of Neuroscience Nursing*, 49(3), 136–136. DOI: 10.1097/JNN.0000000000000288. 10

Optimal Aging Portal Blog Team. (2018). Older adults increasingly targeted by fraud and scams. McMaster University. https://www.mcmasteroptimalaging.org/blog/detail/blog/2018/03/19/older-adults-increasingly-targeted-by-fraud-and-scams. 90

Orpwood, R., Sixsmith, A., Torrington, J., Chadd, J., and Chalfont, G. (2007). Designing technology to support quality of life of people with dementia. *Technology and Disability*, 19(2–3), 103–112. DOI: 10.3233/tad-2007-192-307. 119

Osawa, Y. and Miyazaki, K. (2006). An empirical analysis of the valley of death: Large-scale R&D project performance in a Japanese diversified company. *Asian Journal of Technology Innovation*, 14(2), 93–116. DOI: 10.1080/19761597.2006.9668620. 150

Osteoporosis Canada. (n.d.). Video series on exercise and osteoporosis. https://osteoporosis.ca/video-series-on-exercise-and-osteoporosis/. 103

Ouellet-Léveillé, B. and Milan, A. (2019). Results from the 2016 Census: Occupations with older workers. *Statistics Canada*. https://www150.statcan.gc.ca/n1/pub/75-006-x/2019001/article/00011-eng.htm. 107

Oxford Learner's Dictionary. (2021). *Autonomy*. https://www.oxfordlearnersdictionaries.com/definition/english/autonomy. 5, 173

Ozdemir, D., Cibulka, J., Stepankova, O., and Holmerova, I. (2021). Design and implementation framework of social assistive robotics for people with dementia - a scoping review. *Health and Technology*, 11(2), 367–378. DOI: 10.1007/s12553-021-00522-0. 128

Pacheco, T. B. F., De Medeiros, C. S. P., De Oliveira, V. H. B., Vieira, E. R., and De Cavalcanti, F. A. C. (2020). Effectiveness of exergames for improving mobility and balance in older adults: A systematic review and meta-analysis. *Systematic Reviews*, 9(1), 1–14. DOI: 10.1186/s13643-020-01421-7. 103, 104

Paggi, M. E., Jopp, D., and Hertzog, C. (2016). The importance of leisure activities in the relationship between physical health and well-being in a life span sample. *Gerontology*, 62(4), 450–458. DOI: 10.1159/000444415. 118

Panico, F., Cordasco, G., Vogel, C., Trojano, L., and Esposito, A. (2020). Ethical issues in assistive ambient living technologies for ageing well. *Multimedia Tools and Applications*, 79, 36077–36089. DOI: 10.1007/s11042-020-09313-7. 6

Panizzolo, F. A., Bolgiani, C., Di Liddo, L., Annese, E., and Marcolin, G. (2019). Reducing the energy cost of walking in older adults using a passive hip flexion device. *Journal of NeuroEngineering and Rehabilitation*, 16(1), 1–9. DOI: 10.1186/s12984-019-0599-4. 124

Park, J. Y. E., Li, J., Howren, A., Tsao, N. W., and de Vera, M. (2019). Mobile phone apps targeting medication adherence: Quality assessment and content analysis of user reviews. *JMIR MHealth and UHealth*, 7(1), e11919. DOI: 10.2196/11919. 87

Paro. (2014). PARO therapeutic robot. http://www.parorobots.com/. 128

Patel, M. S., Asch, D. A., and Volpp, K. G. (2015). Wearable devices as facilitators, not drivers, of health behavior change. *JAMA - Journal of the American Medical Association*, 313(5), 459–460. DOI: 10.1001/jama.2014.14781. 100

Patomella, A. H., Lovarini, M., Lindqvist, E., Kottorp, A., and Nygård, L. (2018). Technology use to improve everyday occupations in older persons with mild dementia or mild cognitive impairment: A scoping review. *British Journal of Occupational Therapy*, 81(10), 555–565. DOI: 10.1177/0308022618771533. 87, 165

Patterson, I., Pegg, S., and Litster, J. (2011). Grey nomads on tour: A revolution in travel and tourism for older adults. *Tourism Analysis*, 16(3), 283–294. DOI: 10.3727/108354211X13110944387086. 133

Paul, L. A. (2016). Transformative Experience. Oxford University Press. DOI: 10.1093/acprof:oso/9780198717959.001.0001. 58

Payne, L., Harris, P., Ghio, D., Slodkowska-Barabasz, J., Sutcliffe, M., Kelly, J., Stroud, M., Little, P., Yardley, L., and Morrison, L. (2020). Beliefs about inevitable decline among home-living older adults at risk of malnutrition: A qualitative study. *Journal of Human Nutrition and Dietetics*, 33(6), 841–851. DOI: 10.1111/jhn.12807. 89

Pekkarinen, S., Kuosmanen, P., Melkas, H., Karisto, A., Valve, R., and Kempas, K. (2013). Roles and functions of user-oriented gerontechnology: mStick and hStick. *Journal of Medical and Biological Engineering*, 33(4), 349–355. DOI: 10.5405/jmbe.1291. 100

Pentikäinen, E., Pitkäniemi, A., Siponkoski, S.-T., Jansson, M., Louhivuori, J., Johnson, J. K., Paajanen, T., and Särkämö, T. (2021). Beneficial effects of choir singing on cognition and well-being of older adults: Evidence from a cross-sectional study. *PLOS ONE*, 16(2), e0245666. DOI: 10.1371/journal.pone.0245666. 132

Percival, J., Hanson, J., and Osipovic, D. (2008). Perspectives on telecare: Implications for autonomy, support and social inclusion. *Digital Welfare for the Third Age: Health and Social Care Informatics for Older People*, Routledge. 78

Perilli, V., Lancioni, G. E., Hoogeveen, F., Caffó, A., Singh, N., Sigafoos, J., Cassano, G., and Oliva, D. (2013). Video prompting versus other instruction strategies for persons with Alzheimer's disease. American *Journal of Alzheimer's Disease & Other Dementias*, 28(4), 393–402. DOI: 10.1177/1533317513488913. 90

Petrecca, L. (2019). Free online classes for the masses. AARP. https://www.aarp.org/home-family/personal-technology/info-2019/mooc-online-classes.html. 121

Petrie, H., Carmien, S., and Lewis, A. (2018) Assistive technology abandonment: Research realities and potentials. *Lecture Notes in Computer Science* (including subseries *Lecture Notes in Artificial Intelligence* and *Lecture Notes in Bioinformatics*), pp. 532–540. *2018 16th International Conference on Computers Helping People with Special Needs, ICCHP 2018,* Linz July 11–July 13. 145

Phillipson, C. (2013). Commentary: The future of work and retirement. *Human Relations*, 66(1), 143–153. DOI: 10.1177/0018726712465453. 107

Piercy, K. L., Troiano, R. P., Ballard, R. M., Carlson, S. A., Fulton, J. E., Galuska, D. A., George, S. M., and Olson, R. D. (2018). The physical activity guidelines for Americans. *JAMA*, 320(19), 2020–2028. DOI: 10.1001/jama.2018.14854. 103

PilloHealth. (2019). Pillo health. https://pillohealth.com/. 88

Pivotell. (2016). Pivotell. https://www.pivotell.co.uk/. 88

Polatajko, H. J., Townsend, E. A., and Craik, J. (2013). The Canadian model of occupational performance and engagement (CMOP-E). In *Enabling Occupation II: Advancing an Occupational Therapy Vision of Health, Well-being,and Justice through Occupation* (2nd ed., p. 23). CAOT Publications ACE. 161, 162

Polatajko, H. J., Townsend, E. A., and Whiteford, G. (2007). Canadian model of occupational performance and engagement (CMOP-E). In Townsend, E. A. and Polatajko, H. J., Eds.,

Enabling Occupation II: Advancing an Occupational Therapy Vision of Health, Well-being, and Justice through Occupation. CAOT Publications ACE, pp. 22–36. 83, 173

Poli, A., Kelfve, S., Klompstra, L., Strömberg, A., Jaarsma, T., and Motel-Klingebiel, A. (2020). Prediction of (non)participation of older people in digital health research: Exergame intervention study. *Journal of Medical Internet Research*, 22(6): e17884. DOI: 10.2196/17884. 120

Polyakov, E. V., Mazhanov, M. S., Rolich, A. Y., Voskov, L. S., Kachalova, M. V., and Polyakov, S. V. (2018). Investigation and development of the intelligent voice assistant for the Internet of Things using machine learning. *2018 Moscow Workshop on Electronic and Networking Technologies* (MWENT), 1–5. DOI: 10.1109/MWENT.2018.8337236. 177

Preusse, K. C., Mitzner, T. L., Fausset, C. B., and Rogers, W. A. (2017). Older adults' acceptance of activity trackers. *Journal of Applied Gerontology*, 36(2), 127–155. DOI: 10.1177/0733464815624151. 101

Proffitt, R., Lange, B., Chen, C., and Winstein, C. (2015). A comparison of older adults' subjective experiences with virtual and real environments during dynamic balance activities. *Journal of Aging and Physical Activity*, 23(1), 24–33. DOI: 10.1123/JAPA.2013-0126. 104

Proulx, C. M., Curl, A. L., and Ermer, A. E. (2018). Longitudinal associations between formal volunteering and cognitive functioning. *The Journals of Gerontology: Series B*, 73(3), 522–531. DOI: 10.1093/geronb/gbx110. 110

Pruchno, R. (2019). Technology and aging: An evolving partnership. *The Gerontologist*, 59(1), 1–5. DOI: 10.1093/GERONT/GNY153. xxi, 173

Qato, D. M., Alexander, G. C., Conti, R. M., Johnson, M., Schumm, P., and Lindau, S. T. (2008). Use of prescription and over-the-counter medications and dietary supplements among older adults in the United States. *JAMA - Journal of the American Medical Association*, 300(24), 2867–2878. DOI: 10.1001/jama.2008.892. 87

Quan-Haase, A. and Ho, D. (2020). Online privacy concerns and privacy protection strategies among older adults in East York, Canada. *Journal of the Association for Information Science and Technology*, 71(9), 1089–1102. DOI: 10.1002/asi.24364. 99

Raji, I. D., Gebru, T., Mitchell, M., Buolamwini, J., Lee, J., and Denton, E. (2020). Saving face: Investigating the ethical concerns of facial recognition auditing. *Proceedings of the AAAI/ACM Conference on AI, Ethics, and Society*, 7, 145–151. DOI: 10.1145/3375627.3375820. 170

Ramage-Morin, P. L. (2009). Medication use among senior Canadians. *Health Reports*, 20(1), 37–44. https://citeseerx.ist.psu.edu/viewdoc/download?doi=10.1.1.510.5692&rep=rep1&type=pdf. 87

Rassmus-Gröhn, K. and Magnusson, C. (2014). Finding the way home-supporting wayfinding for older users with memory problems. *Proceedings of the 8th Nordic Conference on Human-Computer Interaction*, 247–255. DOI: 10.1145/2639189.2639233. 135

Ratsoy, G. (2016). The roles of Canadian universities in heterogeneous third-age learning: A call for transformation. *Canadian Journal of Higher Education*, 46(1), 76–90. https://eric.ed.gov/?id=EJ1098228. 121

Rawes, E. (2021). Can Alexa call 911? How to set up Alexa for emergencies. *Digital Trends*. https://www.digitaltrends.com/home/can-alexa-call-911/. 93

Regan, M. A., Hallett, C., and Gordon, C. P. (2011). Driver distraction and driver inattention: Definition, relationship and taxonomy. *Accident Analysis and Prevention*, 43(5), 1771–1781. DOI: 10.1016/j.aap.2011.04.008. 140

Registered Nurses' Association of Ontario. (2020). Rainbow nursing interest group (RNIG). Registered Nurses' Association of Ontario. https://chapters-igs.rnao.ca/interestgroup/58/about. 11

Rempala, K., Hornewer, M., Vukov, J., Meda, R., and Khan, S. (2020). Holding on: A community approach to autonomy in dementia. *American Journal of Bioethics*, 20(8), 107–109. DOI: 10.1080/15265161.2020.1781971. 48, 59

Reppou, S. E., Tsardoulias, E. G., Kintsakis, A. M., Symeonidis, A. L., Mitkas, P. A., Psomopoulos, F. E., Karagiannis, G. T., Zielinski, C., Prunet, V., Merlet, J. P., Iturburu, M., and Gkiokas, A. (2016). Rapp: A robotic-oriented ecosystem for delivering smart user empowering applications for older people. *International Journal of Social Robotics*, 8(4), 539–552. DOI: 10.1007/s12369-016-0361-z. 127

Rewheel Research. (2020). 4G&5G connectivity competitiveness 2020. http://research.rewheel.fi/downloads/4G_5G_connectivity_competitiveness_2020_PUBLIC_VERSION.pdf. 171

Righi, V., Sayago, S., and Blat, J. (2015). Urban ageing: Technology, agency and community in smarter cities for older people. *Proceedings of the 7th International Conference on Communities and Technologies*, June, 119–128. DOI: 10.1145/2768545.2768552. 114

Ríos-Rincón, A., Miguel Cruz, A., Daum, C., Neubauer, N., Comeau, A., and Liu, L. (2021). Digital storytelling in older adults with typical aging, and with cognitive impairment: A systematic literature review. *Journal of Applied Gerontology*. 2021 May 19: 7334648211015456. DOI: 10.1177/07334648211015456. 41

Rios, A., Miguel Cruz, A., Guarín, M. R., and Caycedo Villarraga, P. S. (2014). What factors are associated with the provision of assistive technologies: The Bogotá D.C. case. *Disability and Rehabilitation: Assistive Technology*, 9(5), 432–444. DOI: 10.3109/17483107.2014.936053. 26

Ritchie, H. and Roser, M. (2019). Age structure. *Our World in Data*. https://ourworldindata.org/age-structure. 8

Robertson, R. D., Woods-Fry, H., Hing, M. M., and Vanlaar, W. G. M. (2018). Senior drivers & automated vehicles: Knowledge, attitudes & practices. www.tirf.ca. 141

Robinson, L., Hutchings, D., Corner, L., Finch, T., Hughes, J., Brittain, K., and Bond, J. (2007). Balancing rights and risks: Conflicting perspectives in the management of wandering in dementia. *Health, Risk and Society*, 9(4), 389–406. DOI: 10.1080/13698570701612774. 68, 71

Rogers, E. M. (1995). Diffusion of innovations: Modifications of a model for telecommunications. In Stoetzer, MW. and Mahler, A., Eds., *Die Diffusion von Innovationen in der Telekommunikation. Schriftenreihe des Wissenschaftlichen Instituts für Kommunikationsdienste,* Berlin, Heidelberg: Springer, vol 17, pp. 25-38. DOI: 10.1007/978-3-642-79868-9_2. 146, 177

Rogers, W. A., Meyer, B., Walker, N., and Fisk, A. D. (1998). Functional limitations to daily living tasks in the aged: A focus group analysis. *Human Factors*, 40(1), 111–125. DOI: 10.1518/001872098779480613. 117, 174

Rogers, W. A. and Mitzner, T. L. (2017). Envisioning the future for older adults: Autonomy, health, well-being, and social connectedness with technology support. *Futures*, 87, 133–139. DOI: 10.1016/j.futures.2016.07.002. 56, 84

Rogers, W. A., Mitzner, T. L., and Bixter, M. T. (2020). Understanding the potential of technology to support enhanced activities of daily living (EADLs). *Gerontechnology*, 19(2), 125–137. DOI: 10.4017/gt.2020.19.2.005.00. 101, 117, 139, 141, 174

Rote, S., Hill, T. D., and Ellison, C. G. (2013). Religious attendance and loneliness in later life. *The Gerontologist*, 53(1), 39–50. DOI: 10.1093/geront/gns063. 129

Rush, K. L., Murphy, M. A., and Kozak, J. F. (2012). A photovoice study of older adults' conceptualizations of risk. *Journal of Aging Studies*, 26(4), 448–458. DOI: 10.1016/j.jaging.2012.06.004. 70

Russell, C., Campbell, A., and Hughes, I. (2008). Ageing, social capital and the Internet: Findings from an exploratory study of Australian 'silver surfers.' *Australasian Journal on Ageing*, 27(2), 78–82. DOI: 10.1111/j.1741-6612.2008.00284.x. 127

SAE International. (2018). Taxonomy and Definitions for Terms Related to Driving Automation Systems for On-Road Motor Vehicles (SAE International Patent). DOI: 10.4271/J3016_201806. 140

Safeland. (2021). Safeland - The community collaboration app! https://www.safe.land/gb/. 138

Same, A., McBride, H., Liddelow, C., Mullan, B., and Harris, C. (2020). Motivations for volunteering time with older adults: A qualitative study. *PLOS ONE*, 15(5), e0232718. DOI: 10.1371/journal.pone.0232718. 110

Sampalli, A., Lee, J., and Rizek, P. (2020). Clinical and kinematic assessment of a glove-based tremor dampener in patients with essential tremor. *International Parkinson and Movement Disorder Society*. (35)S1. https://www.mdsabstracts.org/abstract/clinical-and-kinematic-assessment-of-a-glove-based-tremor-dampener-in-patients-with-essential-tremor/. 84

Sánchez, V. G., Taylor, I., and Bing-Jonsson, P. C. (2017). Ethics of smart house welfare technology for older adults: A systematic literature review. *International Journal of Technology Assessment in Health Care*, 33(6), 691–699. DOI: 10.1017/S0266462317000964. 78

Sandry, E. (2015). *Robots and Communication*. Palgrave Macmillan Limited. 127, 177

Sankalpani, P., Wijesinghe, I., Jeewani, I., Anooj, R., Mahadikaara, M. D. J. T. H., and Anuradha Jayakody, J. A. D. C. (2018). "Smart Assistant": A solution to facilitate vision impaired individuals. *2018 National Information Technology Conference (NITC)*, pp. 1–6. DOI: 10.1109/NITC.2018.8550059. 93

Sarrica, M., Brondi, S., and Fortunati, L. (2020). How many facets does a "social robot" have? A review of scientific and popular definitions online. *Information Technology and People*, 33(1), 1–21. DOI: 10.1108/ITP-04-2018-0203. 127, 177

Sas, C. (2018). Exploring self-defining memories in old age and their digital cues. *Proceedings of the 2018 Designing Interactive Systems Conference*, 149–161. DOI: 10.1145/3196709.3196767. 40

Schaper, L. (2021). Digital disruption in the health sector. Waterloo Innovation Summit, April 13, 2021. https://www.youtube.com/watch?v=9NscBYDbv6o. 171

Schehl, B. (2020). Outdoor activity among older adults: Exploring the role of informational Internet use. *Educational Gerontology*, 46(1), 36–45. DOI: 10.1080/03601277.2019.1698200. 125

Schmid, L., Manturuk, K., Simpkins, I., Goldwasser, M., and Whitfield, K. E. (2015). Fulfilling the promise: Do MOOCs reach the educationally underserved? *Educational Media International*, 52(2), 116–128. DOI: 10.1080/09523987.2015.1053288. 122

Schneider, M. J., Jagpal, S., Gupta, S., Li, S., and Yu, Y. (2017). Protecting customer privacy when marketing with second-party data. *International Journal of Research in Marketing*, 34(3), 593–603. DOI: 10.1016/j.ijresmar.2017.02.003. 74

Schrader, L., Vargas Toro, A., Konietzny, S., Rüping, S., Schäpers, B., Steinböck, M., Krewer, C., Müller, F., Güttler, J., and Bock, T. (2020). Advanced sensing and human activity recognition in early intervention and rehabilitation of elderly people. *Journal of Population Ageing* 2020 13:2, 13(2), 139–165. DOI: 10.1007/S12062-020-09260-Z. 132

Schröder-Bäck, P., Duncan, P., Sherlaw, W., Brall, C., and Czabanowska, K. (2014). Teaching seven principles for public health ethics: Towards a curriculum for a short course on ethics in public health programmes. *BMC Medical Ethics*, 15(1), 73. DOI: 10.1186/1472-6939-15-73. 5, 44

Schumacher, S. and Kent, N. (2020). 8 charts on internet use around the world as countries grapple with COVID-19. https://www.pewresearch.org/fact-tank/2020/04/02/8-charts-on-internet-use-around-the-world-as-countries-grapple-with-covid-19/. In Pew Research Center. 133

Seniors First BC. (2021). Assessing legal capacity. http://seniorsfirstbc.ca/for-professionals/assessing-legal-capacity/. 44

Serrat, R., Scharf, T., Villar, F., and Gómez, C. (2019). Fifty-five years of research into older people's civic participation: Recent trends, future directions. *The Gerontologist*, 60(1), E38–E51. DOI: 10.1093/geront/gnz021. 109, 110

Seymour, W. (2018). How loyal is your Alexa? Imagining a respectful smart assistant. *Extended Abstracts of the 2018 CHI Conference on Conference on Human Factors in Computing Systems*. DOI: 10.1145/3170427.3180289. 177

Shalini, S., Levins, T., Robinson, E. L., Lane, K., Park, G., and Skubic, M. (2019). Development and comparison of customized voice-assistant systems for independent living older adults. *International Conference on Human-Computer Interaction*, 11593, 464–479. DOI: 10.1007/978-3-030-22015-0_36. 93

Shapiro, B. and Baker, C. R. (2001). Information technology and the social construction of information privacy. *Journal of Accounting and Public Policy*, 20(4–5), 295–322. DOI: 10.1016/S0278-4254(01)00037-0. 74

Sharit, J. and Czaja, S. J. (2009). Telework and older workers. In Czaja, S. J. and Sharit, J., Eds., *Aging and Work: Issues and Implications in a Changing Landscape*. Johns Hopkins University Press, pp. 126–143. 108

Shen, Y. (2020). The use of bikearound and impact on well-being in older people living in a nursing home: Experiences of the nursing staff. Master's thesis, Halmstad University. http://urn.kb.se/resolve?urn=urn:nbn:se:hh:diva-41722. 124

Sherwin, S. and Winsby, M. (2011). A relational perspective on autonomy for older adults residing in nursing homes. *Health Expectations*, 14(2), 182–190. DOI: 10.1111/j.1369-7625.2010.00638.x. 31

Shishehgar, M., Kerr, D., and Blake, J. (2018). A systematic review of research into how robotic technology can help older people. *Smart Health*, 7–8, 1–18. DOI: 10.1016/j.smhl.2018.03.002. 87, 88, 90

Simhon, Y. and Trites, S. (2017). Financial literacy and retirement well-being in Canada: An analysis of the 2014 Canadian financial capability survey. Financial Consumer Agency of Canada= Agence de la consommation en matière financière du Canada. https://www.canada.ca/content/dam/fcac-acfc/documents/programs/research-surveys-studies-reports/financial-literacy-retirement-well-being.pdf. 90

Simpson, R. C. and Levine, S. P. (2002). Voice control of a powered wheelchair. *IEEE Transactions on Neural Systems and Rehabilitation Engineering*, 10(2), 122–125. DOI: 10.1109/TNSRE.2002.1031981. 93

Sinha, M. (2013). Portrait of caregivers, 2012. www.statcan.gc.ca. 114

Sinner, J. and Wei Lim, Y. (2021). Co-designing technology with elders: A systematic review. *International Journal of Integrated Care*, 20(3), 42. DOI: 10.5334/ijic.s4042. 147

Sisk, B., Alexander, J., Bodnar, C., Curfman, A., Garber, K., McSwain, S. D., and Perrin, J. M. (2020). Pediatrician attitudes toward and experiences with telehealth use: Results from a national survey. *Academic Pediatrics*, 20(5), 628–635. DOI: 10.1016/j.acap.2020.05.004. 171

Sixsmith, A. J., Gibson, G., Orpwood, R. D., and Torrington, J. M. (2007). Developing a technology 'wish-list' to enhance the quality of life of people with dementia. *Gerontechnology*, 6(1). DOI: 10.4017/gt.2007.06.01.002.00. 119

Sixsmith, A. J., Orpwood, R. D., and Torrington, J. M. (2010). Developing a music player for people with dementia. *Gerontechnology*, 9(3). DOI: 10.4017/gt.2010.09.03.004.00. 119

Sixsmith, A. and Gibson, G. (2007). Music and well-being in dementia. *Ageing and Society*, 127–146. DOI: 10.1017/S0144686X06005228. 119

Sixsmith, A., Sixsmith, J., Fang, M. L., and Horst, B. (2020). AgeTech, cognitive health, and dementia. DOI: 10.2200/s01025ed1v01y202006arh014. xxi, 20, 24

Slovic, P. and Peters, E. (2006). Risk perception and affect. *Current Directions in Psychological Science*, 15(6), 322–325. DOI: 10.1111/j.1467-8721.2006.00461.x. 63

Small, J. A., Geldart, K., Gutman, G. M., and Scott, M. (1998). The discourse of self in dementia. *Ageing and Society*, 18(3), 291–316 https://www.researchgate.net/publication/231980943. 38

Smebye, K. L., Kirkevold, M., and Engedal, K. (2012). How do persons with dementia participate in decision making related to health and daily care? A multi-case study. *BMC Health Services Research*, 12(1), 1–12. DOI: 10.1186/1472-6963-12-241. 72

Smebye, K. L., Kirkevold, M., and Engedal, K. (2016). Ethical dilemmas concerning autonomy when persons with dementia wish to live at home: A qualitative, hermeneutic study healthcare needs and demand. *BMC Health Services Research*, 16(21). DOI: 10.1186/s12913-015-1217-1. 31

Smith, A. and Anderson, M. (2018). Social media use in 2018. In Pew Research Center. https://www.pewresearch.org/internet/2018/03/01/social-media-use-in-2018/. 126

Smith, R. O. (2017). Technology and occupation: Past, present, and the next 100 years of theory and practice. *American Journal of Occupational Therapy*, 71(6), 7106150010p1-7106150010p15. DOI: 10.5014/ajot.2017.716003. 21

Smith, R. O., Scherer, M. J., Cooper, R., Bell, D., Hobbs, D. A., Pettersson, C., Seymour, N., Borg, J., Johnson, M. J., Lane, J. P., Sujatha, S., Rao, P., Obiedat, Q. M., MacLachlan, M., and Bauer, S. (2018). Assistive technology products: A position paper from the first global research, innovation, and education on assistive technology (GREAT) summit. *Disability and Rehabilitation: Assistive Technology*, 13(5), 473–485. DOI: 10.1080/17483107.2018.1473895. 21, 24

Sneed, R. S. and Cohen, S. (2013). A prospective study of volunteerism and hypertension risk in older adults. *Psychology and Aging*, 28(2), 578–586. DOI: 10.1037/a0032718. 110

Snyder, S., Hazelett, S., Allen, K., and Radwany, S. (2013). Physician knowledge, attitude, and experience with advance care planning, palliative care, and hospice: Results of a primary care survey. *American Journal of Hospice and Palliative Medicine*, 30(5), 419–424. DOI: 10.1177/1049909112452467. 51

Söderhamn, U., Dale, B., and Söderhamn, O. (2013). The meaning of actualization of self-care resources among a group of older home-dwelling people—A hermeneutic study. *International Journal of Qualitative Studies on Health and Well-Being*, 8(1), 1–9. DOI: 10.3402/qhw.v8i0.20592. 83

Soh, P. Y., Heng, H. B., Selvachandran, G., Anh, L. Q., Chau, H. T. M., Son, L. H., Abdel-Baset, M., Manogaran, G., and Varatharajan, R. (2020). Perception, acceptance and willingness of older adults in Malaysia towards online shopping: A study using the UTAUT and IRT models. *Journal of Ambient Intelligence and Humanized Computing*, 1–13. DOI: 10.1007/s12652-020-01718-4. 90

Sokolowski, M. (2018). *Dementia and the Advance Directive: Lessons from the Bedside*. Springer International Publishing. DOI: 10.1007/978-3-319-72083-8. 58

Son, J. S., Nimrod, G., West, S. T., Janke, M. C., Liechty, T., and Naar, J. J. (2021). Promoting older adults' physical activity and social well-being during COVID-19. *Leisure Sciences*, 43(1–2), 287–294. DOI: 10.1080/01490400.2020.1774015. 122, 133, 134

Sovrin Guardianship Task Force. (2019). On guardianship in self-sovereign identity. https://sovrin.org/wp-content/uploads/Guardianship-Whitepaper2.pdf. 54

Speechmatics. (2020). Barriers to voice technology adoption barriers 2020. *Statista*. https://www-statista-com.proxy.lib.uwaterloo.ca/statistics/1134244/barriers-to-voice-technology-adoption-worldwide/. 93

Starkhammar, S. and Nygård, L. (2008). Using a timer device for the stove: Experiences of older adults with memory impairment or dementia and their families. *Technology and Disability*, 20(3), 179–191. DOI: 10.3233/tad-2008-20302. 89

Statista. (n.d.). Smart home. https://www.statista.com/outlook/dmo/smart-home/worldwide. 91

Statistics Canada. (2018a). A demographic, employment and income profile of Canadians with disabilities aged 15 years and over, 2017. https://www150.statcan.gc.ca/n1/pub/89-654-x/89-654-x2018002-eng.htm. xxi

Statistics Canada. (2018b). Canadian survey on disability, 2017. https://www150.statcan.gc.ca/n1/daily-quotidien/181128/dq181128a-eng.htm. 11, 14

Statistics Canada. (2018c). Type of disability for persons with disabilities aged 15 years and over, by age group and sex, Canada, provinces and territories. xxi

Statistics Canada. (2019). Study: Evolving internet use among Canadian seniors. Statistics Canada. https://www150.statcan.gc.ca/t1/tbl1/en/cv!recreate.action?pid=1310037601&selectedNodeIds=&checkedLevels=0D1,1D1,1D2,2D1,2D2,3D1,3D2,4D1&refPeriods=20170101,20170101&dimensionLayouts=layout2,layout3,layout3,layout3,layout2,layout2&vectorDisplay=false. 165

Statistics Canada. (2020a). Persons with disabilities and COVID-19. Statistics Canada. https://www150.statcan.gc.ca/n1/pub/11-627-m/11-627-m2020040-eng.htm. 8

Statistics Canada. (2020b). The Daily—Industrial research and development characteristics, 2018 (Actual), 2019 (Preliminary) and 2020 (Intentions). Statistics Canada. https://www150. statcan.gc.ca/n1/daily-quotidien/201209/dq201209b-eng.htm. 9, 14

Stenhouse, R., Tait, J., Hardy, P., and Sumner, T. (2013). Dangling conversations: Reflections on the process of creating digital stories during a workshop with people with early-stage dementia. *Journal of Psychiatric and Mental Health Nursing*, 20(2), 134–141. DOI: 10.1111/j.1365-2850.2012.01900.x. 41

Stevens, E. R., Shelley, D., and Boden-Albala, B. (2020). Barriers to engagement in implementation science research: A national survey. *Translational Behavioral Medicine*, 11(2), 408–418 DOI:. 10.1093/tbm/ibz193. 151

Stone, S. D. (2003). Disability, dependence, and old age: Problematic constructions. *Canadian Journal on Aging*, 22(1), 59–67. DOI: 10.1017/S0714980800003731. 8, 9

Straub, E. T. (2009). Understanding technology adoption: Theory and future directions for informal learning. *Review of Educational Research*, 79(2), 625–649. DOI: 10.3102/0034654308325896. 146, 177

Stuck, R. E., Chong, A. W., Mitzner, T. L., and Rogers, W. A. (2017). Medication management apps: Usable by older adults? *Proceedings of the Human Factors and Ergonomics Society ... Annual Meeting. Human Factors and Ergonomics Society. Annual Meeting*, 61(1), pp. 1141–1144. DOI: 10.1177/1541931213601769. 88

Szanton, S. L., Walker, R. K., Roberts, L., Thorpe, R. J., Wolff, J., Agree, E., Roth, D. L., Gitlin, L. N., and Seplaki, C. (2015). Older adults' favorite activities are resoundingly active: Findings from the NHATS study. *Geriatric Nursing*, 36(2), 131–135. DOI: 10.1016/j.gerinurse.2014.12.008. 103, 122

Tabi, K., Randhawa, A. S., Choi, F., Mithani, Z., Albers, F., Schnieder, M., Nikoo, M., Vigo, D., Jang, K., Demlova, R., and Krausz, M. (2019). Mobile apps for medication management: Review and analysis. *JMIR MHealth and UHealth*, 7(9), e13608. DOI: 10.2196/13608. 87, 88

TAGlab. (n.d.-a). Our projects. http://taglab.utoronto.ca/index.php/projects/language-customization-tool-to-simplify-health-information/. 99

TAGlab. (n.d.-b). Improving the on-line safety of older Canadian adults. http://taglab.utoronto.ca/index.php/projects/improving-the-on-line-safety-of-older-canadian-adults-understanding-and-removing-barriers-to-technology-adoption/. 99

Taiebat, M., Brown, A. L., Safford, H. R., Qu, S., and Xu, M. (2018). A review on energy, environmental, and sustainability implications of connected and automated vehicles. *Environ-*

mental Science & Technology, 52(20), 11449–11465. DOI: 10.1021/ACS.EST.8B00127. 141

Tan, J. P., Li, N., Gao, J., Wang, L. N., Zhao, Y. M., Yu, B. C., Du, W., Zhang, W. J., Cui, L. Q., Wang, Q. S., Li, J. J., Yang, J. S., Yu, J. M., Xia, X. N., and Zhou, P. Y. (2015). Optimal cutoff scores for dementia and mild cognitive impairment of the Montreal Cognitive Assessment among elderly and oldest-old Chinese population. *Journal of Alzheimer's disease* (JAD), 43(4), 1403–1412. DOI: 10.3233/JAD-141278. 45

Taylor, J. E., Connolly, M. J., Brookland, R., and Samaranayaka, A. (2018). Understanding driving anxiety in older adults. *Maturitas*, 118, 51–55. DOI: 10.1016/j.*maturitas*.2018.10.008. 104, 140

Teipel, S., Babiloni, C., Hoey, J., Kaye, J., Kirste, T., and Burmeister, O. K. (2016). Information and communication technology solutions for outdoor navigation in dementia. *Alzheimer's and Dementia*, 12(6), 695–707. DOI: 10.1016/j.jalz.2015.11.003. 56

The Economist. (2021). At 54, China's average retirement age is too low. https://www.economist.com/china/2021/06/22/chinas-average-retirement-age-is-ridiculously-low-54. 107

The Institute for Successful Longevity. (2020). ISL launches Zoom initiative to help older adults fight social isolation. Florida State University. https://isl.fsu.edu/article/isl-launches-zoom-initiative-help-older-adults-fight-social-isolation. 165

Titterton, M. (2005). *Risk and Risk Taking in Health and Social Welfare*. Jessica Kingsley Publishers: London, UK. 63

Tong, T., Chignell, M., Tierney, M. C., and Lee, J. (2016). A serious game for clinical assessment of cognitive status: Validation study. *JMIR Serious Games*, 4(1), e7. DOI: 10.2196/games.5006. 45

Townsend, D., Knoefel, F., and Goubran, R. (2011). Privacy versus autonomy: A tradeoff model for smart home monitoring technologies. *2011 Annual International Conference of the IEEE Engineering in Medicine and Biology Society*, 4749–4752. DOI: 10.1109/IEMBS.2011.6091176. 169

Trajkova, M. and Martin-Hammond, A. (2020). "Alexa is a toy": Exploring older adults' reasons for using, limiting, and abandoning echo. *Proceedings of the 2020 CHI Conference on Human Factors in Computing Systems*, 1–13. DOI: 10.1145/3313831.3376760. 177

Tricella. (2020). Tricella. https://www.tricella.com/. 88

Truth and Reconciliation Commission of Canada. (2015). Honouring the truth, reconciling for the future: Summary of the final report of the Truth and Reconciliation Commission of Canada. Library and Archives Canada. www.trc.ca. 162

Tyack, C., Camic, P. M., Heron, M. J., and Hulbert, S. (2017). Viewing art on a tablet computer: A well-being intervention for people with dementia and their caregivers. *Journal of Applied Gerontology*, 36(7), 864–894. DOI: 10.1177/0733464815617287. 132

UCLA Mindful Awareness Research Center. (2017). Guided meditations. https://www.uclahealth.org/marc/mindful-meditations. 129

United Nations. (2007). Convention on the rights of persons with disabilities, Resolution adopted by the General Assembly on 13 December 2006 (2007). https://www.un.org/en/development/desa/population/migration/generalassembly/docs/globalcompact/A_RES_61_106.pdf. 25, 26

United Nations. (2016). Ageing and disability. Department of Economic and Social Affairs Disability. https://www.un.org/development/desa/disabilities/disability-and-ageing.html. 8

United Nations. (2021). Ageing and disability. https://www.un.org/development/desa/disabilities/disability-and-ageing.html. xxi

Vailshery, L. S. (2021). Number of voice assistants in use worldwide 2019-2024. *Statista*. https://www.statista.com/statistics/973815/worldwide-digital-voice-assistant-in-use/.

Valladares-Rodriguez, S., Fernández-Iglesias, M. J., Anido-Rifón, L., Facal, D., and Pérez-Rodríguez, R. (2018). Episodix: A serious game to detect cognitive impairment in senior adults. A psychometric study. *PeerJ*, 2018(9), e5478. DOI: 10.7717/peerj.5478. 45

Van Aerschot, L. and Parviainen, J. (2020). Robots responding to care needs? A multitasking care robot pursued for 25 years, available products offer simple entertainment and instrumental assistance. *Ethics and Information Technology*, 22, 247–256. DOI: 10.1007/s10676-020-09536-0. 87

Van Brummelen, J., O'Brien, M., Gruyer, D., and Najjaran, H. (2018). Autonomous vehicle perception: The technology of today and tomorrow. *Transportation Research Part C: Emerging Technologies*, 89, 384–406. DOI: 10.1016/j.trc.2018.02.012. 139

Van Cauwenberg, J., De Bourdeaudhuij, I., Clarys, P., de Geus, B., and Deforche, B. (2019). E-bikes among older adults: Benefits, disadvantages, usage and crash characteristics. *Transportation*, 46(6), 2151–2172. DOI: 10.1007/s11116-018-9919-y. 123

Van der Roest, H. G., Wenborn, J., Pastink, C., Dröes, R. M., and Orrell, M. (2017). Assistive technology for memory support in dementia. *Cochrane Database of Systematic Reviews*, 6. DOI: 10.1002/14651858.CD009627.pub2. 87

van Deursen, A. J. and Helsper, E. J. (2015). A nuanced understanding of Internet use and non-use among the elderly. *European Journal of Communication*, 30(2), 171–187. DOI: 10.1177/0267323115578059. 126

Vasconcelos, A. F. (2018). Older workers as a source of wisdom capital: Broadening perspectives. *Revista de Gestão*, 25(1), 102–118. DOI: 10.1108/REGE-11-2017-002. 109

Venkatesh, V., Thong, J. Y. L., and Xu, X. (2012). Consumer acceptance and use of information technology: Extending the unified theory of acceptance and use of technology. *MIS Quarterly: Management Information Systems*, 36(1), 157–178. DOI: 10.2307/41410412. 145

Vézina, M. and Crompton, S. (2012). Volunteering in Canada. https://www150.statcan.gc.ca/n1/pub/11-008-x/2012001/article/11638-eng.pdf. 110

Vezzoni, C. (2005). The legal status and social practice of treatment directives in the Netherlands. Doctoral dissertation, University of Groningen. https://research.rug.nl/en/publications/the-legal-status-and-social-practice-of-treatment-directives-in-t. 58

Vital Tracker. (n.d.). Home care monitoring. https://vitaltracer.com/home-care-monitoring/. 100

Voicebot.ai and Business Wire. (2021). Number of voice assistants in use worldwide 2019-2024 (in billions). *Statista*. https://www-statista-com.proxy.lib.uwaterloo.ca/statistics/973815/worldwide-digital-voice-assistant-in-use/.

von Humboldt, S., Mendoza-Ruvalcaba, N. M., Arias-Merino, E. D., Costa, A., Cabras, E., Low, G., and Leal, I. (2020). Smart technology and the meaning in life of older adults during the Covid-19 public health emergency period: A cross-cultural qualitative study. *International Review of Psychiatry*, 32(7–8), 713–722. DOI: 10.1080/09540261.2020.1810643. 98, 103, 126, 129

Wall Communications Inc. (2019). Price comparisons of wireline, wireless and internet services in Canada and with foreign jurisdictions. https://www.ic.gc.ca/eic/site/693.nsf/vwapj/2019_Pricing_Study_Report_EN.pdf/$FILE/2019_Pricing_Study_Report_EN.pdf 171

Wallace, B., Knoefel, F., Goubran, R., Masson, P., Baker, A., Allard, B., Stroulia, E., and Guana, V. (2017). Monitoring cognitive ability in patients with moderate dementia using a modified "whack-a-mole." *2017 IEEE International Symposium on Medical Measurements and Applications, MeMeA 2017 - Proceedings*, 292–297. DOI: 10.1109/MeMeA.2017.7985891. 45

Walsh, E. (2020). Cognitive transformation, dementia, and the moral weight of advance directives. *American Journal of Bioethics*, 20(8), 54–64. DOI: 10.1080/15265161.2020.1781955. 51, 58, 173

Wang, S., Bolling, K., Mao, W., Reichstadt, J., Jeste, D., Kim, H.-C., and Nebeker, C. (2019). Technology to support aging in place: Older adults' perspectives. *Healthcare*, 7(2), 60. DOI: 10.3390/healthcare7020060. 75

Watkins, I. and Xie, B. (2014). eHealth literacy interventions for older adults: A systematic review of the literature. *Journal of Medical Internet Research*, 16(11), e225. DOI: 10.2196/jmir.3318. 98

Waycott, J., Davis, H., Warr, D., Edmonds, F., and Taylor, G. (2017). Co-constructing meaning and negotiating participation: Ethical tensions when "giving voice" through digital storytelling. *Interacting with Computers*, 29(2), 237–247. DOI: 10.1093/iwc/iww025. 41

West, D., Quigley, A., and Kay, J. (2007). MEMENTO: A digital-physical scrapbook for memory sharing. *Personal and Ubiquitous Computing*, 11, 313–328. DOI: 10.1007/s00779-006-0090-7. 41

Westerhof, G. J., Whitbourne, S. K., and Freeman, G. P. (2012). The aging self in a cultural context: The relation of conceptions of aging to identity processes and self-esteem in the United States and the Netherlands. *The Journals of Gerontology: Series B*, 67B(1), 52–60. DOI: 10.1093/geronb/gbr075. 37

Westin, A. F. (1967). *Privacy and Freedom*, 7th ed. *Washington and Lee Law Review*. 73

Whitbourne, S. K., Sneed, J. R., and Skultety, K. M. (2002). Identity processes in adulthood: Theoretical and methodological challenges. *Identity*, 2(1), 29–45. DOI: 10.1207/s1532706xid0201_03. 37

Whitehead, P. J., Drummond, A. E. E., Walker, M. F., and Parry, R. H. (2013). Interventions to reduce dependency in personal activities of daily living in community-dwelling adults who use homecare services: Protocol for a systematic review. *Systematic Reviews*, 2(1), 1–7. DOI: 10.1186/2046-4053-2-49. 83, 173

Whitelock, E. and Ensaff, H. (2018). On your own: Older adults' food choice and dietary habits. *Nutrients*, 10(4), 413. DOI: 10.3390/nu10040413. 88, 89

Wickins, A., Daum, C., Ríos-Rincón, A., Miguel-Cruz, A., and Liu, L. (2019). Flow in older adults: Observations as an alternative to self-report. *Canadian Association of Occupational Therapists Conference 2019*. https://caot.ca/client/relation_roster/clientRelationRosterView.html?clientRelationRosterId=80&language=fr_FR&page=2. 131

Wiemeyer, J. and Kliem, A. (2012). Serious games in prevention and rehabilitation-a new panacea for elderly people? *European Review of Aging and Physical Activity*, 9(1), 41–50. DOI: 10.1007/s11556-011-0093-x. 104

Wiener, J. M., Hanley, R. J., Clark, R., and Van Nostrand, J. F. (1990). Measuring the activities of daily living: Comparisons across national surveys. *Journal of Gerontology*, 45(6), S229–S237. https://academic.oup.com/geronj/article/45/6/S229/706347. 86

Wilińska, M., de Hontheim, A., and Anbäcken, E.-M. (2018). Ageism in a cross-cultural perspective: Reflections from the research field. *International Perspectives on Aging* 19, pp. 425–440. DOI: 10.1007/978-3-319-73820-8_26. 37

Wilkinson, P. (2016). A brief history of serious games. In Dörner, R., Göbel, S., Kickmeier-Rust, M., Masuch, M., and Zweig, K., Eds., *Entertainment Computing and Serious Games*. Switzerland: Springer International Publishing, pp. 17–41. 45

Wolfe, S. E., Greenhill, B., Butchard, S., and Day, J. (2020). The meaning of autonomy when living with dementia: A Q-method investigation. *Dementia,* 20(6), pp. 1875–1890. DOI: 10.1177/1471301220973067. 48, 50

Wong, C. K., Mentis, H. M., and Kuber, R. (2018). The bit doesn't fit: Evaluation of a commercial activity-tracker at slower walking speeds. *Gait & Posture*, 59, 177–181. DOI: 10.1016/J.GAITPOST.2017.10.010. 100

World Health Organization. (2002). Towards a common language for functioning, disability and health: ICF. World Health Organization. https://www.who.int/docs/default-source/classification/icf/icfbeginnersguide.pdf?sfvrsn=eead63d3_4. 18, 173, 174, 176

World Health Organization (2015). World report on aging and health. http://apps.who.int/iris/bitstream/handle/10665/186463/9789240694811_eng.pdf;jsessionid=84654849609918B-D025AD2A4ECC584A6?sequence=1. 79

World Health Organization. (2016). Priority assistive products list. https://www.who.int/phi/implementation/assistive_technology/global_survey-apl/en/. 26

World Health Organization. (2017). Mental health of older adults. https://www.who.int/news-room/fact-sheets/detail/mental-health-of-older-adults. xxi

World Health Organization. (2020). Disability and health. https://www.who.int/news-room/fact-sheets/detail/disability-and-health. xxi

World Health Organization. (2021). Nurses by sex (%). The Global Health Observatory. https://www.who.int/data/gho/data/indicators/indicator-details/GHO/nurses-by-sex-(-). 11

Wright, M. S. (2020). Dementia, autonomy, and supported healthcare decisionmaking. *Maryland Law Review*, 79(2), 257–324. https://elibrary.law.psu.edu/fac_works/389. 6

Wu, Y. H., Wrobel, J., Cornuet, M., Kerhervé, H., Damnée, S., and Rigaud, A. S. (2014). Acceptance of an assistive robot in older adults: A mixed-method study of human-robot interaction over a 1-month period in the living lab setting. *Clinical Interventions in Aging*, 9, 801–811. DOI: 10.2147/CIA.S56435. 40

Xie, B., Charness, N., Fingerman, K., Kaye, J., Kim, M. T., and Khurshid, A. (2020). When going digital becomes a necessity: Ensuring older adults' needs for information, services, and social inclusion during COVID-19. *Journal of Aging & Social Policy*, 32(4–5), 460–470 DOI:. 10.1080/08959420.2020.1771237. 100, 165

Xie, B., Watkins, I., Golbeck, J., and Huang, M. (2012). Understanding and changing older adults' perceptions and learning of social media. *Educational Gerontology*, 38(4), 282–296. DOI: 10.1080/03601277.2010.544580. 100

Yatawara, C., Lee, D. R., Lim, L., Zhou, J., and Kandiah, N. (2017). Getting lost behavior in patients with mild Alzheimer's disease: A cognitive and anatomical model. *Frontiers in Medicine*, 4, 16. DOI: 10.3389/fmed.2017.00201. 135

Yates, L. A., Ziser, S., Spector, A., and Orrell, M. (2016). Cognitive leisure activities and future risk of cognitive impairment and dementia: Systematic review and meta-analysis. *International Psychogeriatrics*, 28(11), 1791–1806. DOI: 10.1017/S1041610216001137. 118

Zheng, L., Li, G., Wang, X., Yin, H., Jia, Y., Leng, M., Li, H., and Chen, L. (2019). Effect of exergames on physical outcomes in frail elderly: A systematic review. *Aging Clinical and Experimental Research*, 32(11), 2187–2200. DOI: 10.1007/s40520-019-01344-x. 103, 104, 174

Zlatintsi, A., Dometios, A. C., Kardaris, N., Rodomagoulakis, I., Koutras, P., Papageorgiou, X., Maragos, P., Tzafestas, C. S., Vartholomeos, P., Hauer, K., Werner, C., Annicchiarico, R., Lombardi, M. G., Adriano, F., Asfour, T., Sabatini, A. M., Laschi, C., Cianchetti, M., Güler, A., … López, R. (2020). I-Support: A robotic platform of an assistive bathing robot for the elderly population. *Robotics and Autonomous Systems*, 126, 103451. DOI: 10.1016/j.robot.2020.103451. 87

Authors' Biographies

Lili Liu earned B.Sc. (Occupational Therapy), M.Sc. and Ph.D. (Rehabilitation Science) degrees at McGill University, Montreal, Canada. Prior to her graduate education, she worked as an occupational therapist in adult mental health which included older adults living with cognitive impairments. Her academic career began at the University of Alberta, Faculty of Rehabilitation Medicine where she served as chair of the Department of Occupational Therapy for over a decade, and where she maintains Adjunct Professor status. Currently, she is Dean of the Faculty of Health at the University of Waterloo, and a Professor in the School of Public Health Sciences. She leads the Aging and Innovation Research Program, with external funding to support undergraduate and graduate students, and postdoctoral fellows who are pursuing careers in aging and health.

Christine Daum earned B.Sc. (Occupational Therapy), M.Sc. (Health Promotion), and Ph.D. (Rehabilitation Science) degrees at the University of Alberta, Edmonton, Canada. She worked as an occupational therapist in community rehabilitation, community care, and long-term care with older adults as well as persons living with brain injuries in Canada and the Cayman Islands. Christine is a Research Assistant Professor and coordinates Lili Liu's Aging and Innovation Research Program at the University of Waterloo, and the University of Alberta. She has the privilege of working closely with older adults, care partners, and community organizations to facilitate research that is relevant to their needs and those of their communities.

Noelannah Neubauer earned B.HK. (Human Kinetics), M.Sc. (Interdisciplinary Studies) degrees at University of British Columbia Okanagan. She also completed M.Sc. in Occupational Therapy and Ph.D. in Rehabilitation Science degrees at the University of Alberta. She is currently a postdoctoral fellow in Lili Liu's Aging and Innovation Research Program at the University of Waterloo. She is the co-founder of the International Consortium on Dementia and Wayfinding and co-founder of OTech Canada.

Antonio Miguel Cruz earned a B.Sc. (Nuclear Engineering) degree at the Nuclear Science Institute, Habana, Cuba, and M.Sc. (Bioengineering) and Ph.D. (Bioengineering) degrees at the Technological University, Habana, Cuba. His academic career began at the Technological University, Faculty of Electrical Engineering, where he was Associate Professor and chair of the Bioengineering Centre. He also served as Full Professor and the Chair of the Biomedical Engineering Program, Universidad del Rosario, School of Medicine and Health Sciences, Bogotá, Colombia. Currently, he is an Assistant Teaching Professor, Department of Occupational Therapy, Faculty of Rehabilitation Medicine, University of Alberta, and a researcher at Glenrose Rehabilitation Research, Innovation & Technology (GRRIT). He is an Adjunct Assistant Professor in the School of Public Health Sciences, Faculty of Health at the University of Waterloo. His research focuses on technology adoption in health.

Adriana Ríos Rincón earned a B.Sc. (Occupational Therapy) at the Universidad Nacional de Colombia, and a M.Sc. (Biomedical Sciences) at the Universidad de Los Andes, both in Bogotá, Colombia. She earned a Ph.D. (Rehabilitation Science) degree at the University of Alberta, Edmonton, Canada. Her academic career began as an Assistant Professor at the Universidad del Rosario, School of Medicine and Health Sciences, Bogotá, Colombia, where she also served as the director of a M.Sc. program in Rehabilitation Science for over a year. Currently, she is an Assistant Professor in the Department of Occupational Therapy, Faculty of Rehabilitation Medicine, University of Alberta. She is an Adjunct Assistant Professor in the School of Public Health Sciences, Faculty of Health at the University of Waterloo. Her research program examines the effects of implementing advanced technologies on the occupational performance, functioning, and social participation of people with disabilities and older adults.

Printed in the United States
by Baker & Taylor Publisher Services